MATERIALS
THERMODYNAMICS

MATERIALS THERMODYNAMICS

Y. Austin Chang
Department of Materials Science and Engineering
University of Wisconsin
Madison, Wisconsin

W. Alan Oates
Institute for Materials Research
University of Salford
Salford, United Kingdom

A JOHN WILEY & SONS, INC., PUBLICATION

Published by John Wiley & Sons, Inc., Hoboken, New Jersey
Published simultaneously in Canada

Limit of Liability/Disclaimer of Warranty: While the publisher and author have used their best efforts in preparing this book, they make no representations or warranties with respect to the accuracy or completeness of the contents of this book and specifically disclaim any implied warranties of merchantability or fitness for a particular purpose. No warranty may be created or extended by sales representatives or written sales materials. The advice and strategies contained herin may not be suitable for your situation. You should consult with a professional where appropriate. Neither the publisher nor author shall be liable for any loss of profit or any other commercial damages, including but not limited to special, incidental, consequential, or other damages.

For general information on our other products and services or for technical support, please contact our Customer Care Department with the United States at (800) 762-2974, outside the United States at (317) 572-3993 or fax (317) 572-4002.

Wiley also publishes its books in a variety of electronic formats. Some content that appears in print may not be available in electronic formats. For more information about Wiley products, visit our web site at www.wiley.com.

Library of Congress Cataloging-in-Publication Data:

Chang, Y. Austin.
  Materials thermodynamics / Y. Austin Chang, W. Alan Oates.
    p. cm. – (Wiley series on processing engineering materials)
  ISBN 978-0-470-48414-2 (cloth)
1. Materials–Thermal properties. 2. Thermodynamics. I. Oates, W. Alan, 1931-II. Title.
  TA418.52.C43 2010
  620.1'1296–dc22
                              2009018590

Printed in the United States of America

10 9 8 7 6 5 4 3 2 1

Thanks to my spouse, Jean Chang, for constant support, and a special tribute to my mother, Shu-Ying Chang, for teaching and guiding me during my youthful days in a remote village in Henan, China

Y. A. C.

Contents

Preface

Much of this book has been used for a course in thermodynamics for beginning graduate students in materials science and engineering (MS&E) and is considered as core material. Those who enroll in the course come with a variety of backgrounds, although all have encountered thermodynamics at least once in their previous studies so that a minimum amount of time is spent on the fundamentals of the subject.

As compared with the available texts on MS&E thermodynamics, we think that the material covered in this book can claim to adopt a more modern approach in that we:

(A) Recognize the impact of the computer on the *teaching* of MS&E thermodynamics. While the impact of computers on the application of thermodynamics in industry is widely known, their influence on the teaching of thermodynamics to MS&E students has not been sufficiently recognized in texts to date. Our philosophy on how computers can best be utilized in the teaching environment is given in more detail below.

(B) Make the students aware of the practical problems in using thermodynamics. It has been our experience that it is easy for students to be seduced by the charming idea of the ability of thermodynamics to predict something from nothing. Many seem to believe that one has only to sit down with a piece of commercial software and request the prediction of equilibrium in the X–Y–Z system. In an effort to enable students to have a more realistic outlook, we have placed a lot of emphasis on system definition. Proper system definition can be particularly difficult when considering chemical equilibria in high-temperature systems. The ability to arrive at incorrect results from thermodynamic calculations on a poorly defined system is something of which all students should be made aware.

(C) Emphasize that the calculation of the position of phase and chemical equilibrium in complex systems, even when properly defined, is not easy. It usually involves finding a constrained minimum in the Gibbs energy. It is nevertheless possible to illustrate the principles involved to students and this we have set out to try and do. With this aim in mind, the use of Lagrangian multipliers is introduced early on for the simplest case of phase equilibrium in unary systems. The same procedure is then followed in its application to phase equilibria in binary systems and the calculation of chemical equilibria in complex systems.

(D) Relegate concepts like equilibrium constants, activities, activity coefficients, free-energy functions, Gibbs–Duhem integrations, all of which seemed so important in the teaching of thermodynamics 50 years ago, to a relatively minor role. This change in emphasis is again a result of the impact of computer-based calculations.

(E) Consider the use of approximations of higher order than the usual Bragg–Williams in solution-phase modeling.

THE ROLE OF COMPUTERS IN THE TEACHING OF THERMODYNAMICS

New tools can lead us to new, better ways of teaching, if only we apply a modicum of creativity and common sense. In the process some of our cherished traditions will need to go, as will some of what we teach now.

There is no denying the impact of computers on the application of thermodynamics to practical problems in MS&E. There is an equal expectation that computers should also have a major influence on the way that thermodynamics is taught to students in this discipline. One might ask whether ground-breakers like Lewis and Randall, who did so much to influence the way that chemical thermodynamics was taught some 80 or so years ago, would have done things as they did if they had had access to a modern computer.

Some consideration is needed, however, to achieve the optimum use of computers in a course such as the one covered by these chapters. There can be no argument about the tremendous graphics capabilities of computers, a tool capable of providing a vast improvement in the presentation of results from thermodynamic calculations. This aspect alone would have delighted Gibbs, who was keen on using plaster models for property visualization.

We firmly believe that the use of commercial blackbox thermodynamic software, widely used in research and industry, has no place in the *teaching* of thermodynamics. It is much more important that students understand the fundamentals of the problem they are trying to solve. As an intermediate way, between the use of blackbox thermodynamic software and students having to write their own programs, we believe that the use of nonlinear equation solvers is to be preferred. Their use demands that students understand the fundamentals and yet still offers the advantages afforded only by the use of computers without requiring the student to be either a skilled programmer or an expert in numerical methods. There are many appropriate packages around which are suitable for this task—Matlab©, MathCad©, EES©, and Solver in Microsoft Excel© are examples. Many of the problems in these chapters are written around the use of such programs. While there remain some of the older style of problem, with simple models, used for hand calculation, their use permits the student to encounter problems based on real systems.

The solution of thermodynamic problems invariably involves the minimization of a function or the solution of a set of nonlinear equations. Before computers, this led to the simplification of problems so that they were made amenable for hand calculation. With computer assistance, however, it is no longer necessary that student exercises involve only perfect gases or ideal or regular solutions or that the calculation of phase diagrams should be confined to the simplest two-phase equilibrium problems. With a computer, solving real system problems becomes just as straightforward. There is the added advantage that computational errors, so common in hand calculations, are more easily avoided.

There is another important aspect to the impact of computers in thermodynamics: They have changed the way in which calculations are carried out and we believe that this should also be reflected in the material presented in a course on MS&E Thermodynamics. For example:

1. In the teaching of phase equilibria, it has been usual to consider the equality of chemical potentials of all components in all phases, $\mu_i^{(i)} = \mu_i^{(j)}$, as the cornerstone of such calculations. Although not denying the importance of students learning the derivation and application of these criteria, they should also appreciate that its application is restricted to calculating the equilibrium between two prespecified phases. When more complex situations have to be considered, a whole new philosophy for carrying out phase equilibria calculations is needed. Modern students should be aware of these developments.

2. The equation $\Delta G^\circ = -RT \log_e K^\circ$ has played a key role in the teaching of chemical thermodynamics, but it should be remembered that this equation is applicable to a single-chemical-reaction equation. Real-world applications, however, often involve systems containing many species and, therefore, many independent reaction equations. The solution of the chemical equilibrium problem for such complex systems requires a different approach. Again, while not denying that students should derive and use the equilibrium constant equation, we believe that they should also be made aware of how more complex systems can be handled through the use of computers.

3. Analytical representation of thermodynamic properties, no matter how complex the function required, is also of little concern when fed into a computer. But it has changed the way that students spend their time in learning the fundamentals. Things like the counting of squares or the weighing of paper cut-outs for the graphical solution of Gibbs–Duhem integrations, so much a part of the learning experience in the precomputer era, are a thing of the past.

In considering both unary and binary systems, the approach has been to present the material in the following order:

  (i) Macroscopic thermodynamics
 (ii) Microscopic models
(iii) Phase equilibria

We are aware of some of the shortcomings of the material presented. Specifically:

1. The book is restricted almost completely to a consideration of unary and binary systems, although there are two chapters specifically on ternary alloys and, in the case of homogeneous chemical equilibria, emphasis is placed on treating multispecies systems.
2. We have avoided using any statistical mechanics in the chapters: There is no mention of Hamiltonians or partition functions. We have also avoided any mention of correlation functions by using the equivalent cluster probabilities. Students who have completed the course should be in good shape to follow up with a more formal study of statistical mechanics. Similarly, the text is restricted to a consideration of alloys only but, again, students who understand this material would have no difficulty in applying the concepts to other types of materials.
3. There is almost nothing on stress as a variable since this requires a specialized background of its own. For an excellent tutorial on this topic see W. C. Johnson, "Influence of Stress on Phase Transformations," in *Lectures on the Theory of Phase Transformations*, Second Edition, H. Aaronson, Editor, The Minerals, Metals and Materials Society, Warrendale, PA, 1999, p. 35.
4. There is almost nothing on the influence of pressure on phase equilibria.

FURTHER READING

The following classic texts are especially valuable in giving an insight into the meaning of thermodynamics and should be consulted by all students wishing to specialize in the subject.

H. B. Callen, *Thermodynamics and an Introduction to Thermostatics*, 2nd ed., Wiley, New York, 1985.

K. G. Denbigh, *Principles of Chemical Equilibrium: With Applications to Chemistry and Chemical Engineering*, 4th ed., Cambridge University Press, 1981.

E. A. Guggenheim, *Thermodynamics*, 5th ed., North-Holland, 1967.

A. P. Pippard, *Classical Thermodynamics*, Cambridge University Press, 1960.

I. Prigogine and R. Defay, *Chemical Thermodynamics*, translated by D. H. Everett, Longmans Green and Co., 1954.

H. Reiss, *Methods of Thermodynamics*, Dover Publications, 1996.

J. W. Tester and M. Modell, *Thermodynamics and its Applications*, Prentice-Hall, 3rd ed., 1996

Some books which are oriented to the application of thermodynamics to MS&E and hence to the material covered in the present text are, in order of year of publication:

C. Wagner, *Thermodynamics of Alloys*, Addison-Wesley, 1952.

L. S. Darken and R. W. Gurry, *Physical Chemistry of Metals*, McGraw-Hill, 1953.

A. Prince, *Alloy Phase Equilibria*, Elsevier, Amsterdam, 1966.

T. B. Reed, *Free Energy of Formation of Binary Compounds*, MIT Press, Boston, 1971.

R. A. Swalin, *Thermodynamics of Solids*, 2nd ed., Wiley, New York, 1972.

E. T. Turkdogan, *Physical Chemistry of High Temperature Technology*, Academic, New York, 1980.

C. H. P. Lupis, *Chemical Thermodynamics of Materials*, North-Holland, 1983.

O. F. Devereux, *Topics in Metallurgical Thermodynamics*, Wiley, New York, 1983.

O. Kubaschewski, C. B. Alcock, and P. J. Spencer, *Materials Thermochemistry*, 6th ed., Pergamon, 1993.

D. R. Gaskell, *Introduction to the Thermodynamics of Materials*, 3rd ed., Taylor & Francis, 1995.

D. V. Ragone, *Thermodynamics of Materials*, Vols. I and II, Wiley, New York, 1995.

D. L. Johnson and G. B. Stracher, *Thermodynamic Loop Applications in Materials Systems*, Minerals, Metals and Materials Society, Warrendale, PA, 1995.

J. B. Hudson, *Thermodynamics of Materials: A Classical and Statistical Synthesis*, Wiley, New York, 1996.

N. G. Saunders and A. P. Miodownik, *CALPHAD: A Comprehensive Guide*, Pergamon Materials Series, Elsevier, Amsterdam, 1996.

M. Hillert, *Phase Equilibria, Phase Diagrams and Phase Transformations: Their Thermodynamic Basis*, Cambridge University Press, 1998.

H. Aaronson, Ed., *Lectures on the Theory of Phase Transformations*, 2nd ed., Minerals, Metals and Materials Society, Warrendale, PA, 1999.

D. R. F. West and N. Saunders, *Ternary Phase Diagrams in Materials Science*, 3rd ed., Institute of Materials, London, 1992.

B. Predel, M. Hoch, and M. Pool, *Phase Diagrams and Heterogeneous Equilibria: A Practical Introduction*, Springer-Verlag, Berlin, 2004.

S. Stolen and T. Grande, *Chemical Thermodynamics of Materials: Macroscopic and Microscopic Aspects*, Wiley, Hoboken, NJ, 2005.

R. T. DeHoff, *Thermodynamics in Materials Science*, 2nd ed., CRC Press, Boca Raton, FL, 2006.

H. L. Lukas, S. G. Fries, and B. Sundman, *Computational Thermodynamics: The Calphad Method*, Cambdidge University Press, 2007.

ACKNOWLEDGMENTS

We both have taught thermodynamics to metallurgy and materials science and engineering students for several decades. Much of the material presented has, of course, drawn heavily on previously published books and articles. The origin of this material has, in many cases, long since been forgotten. To those who might recognize their work and see it unacknowledged, we can only apologize.

To Andy Watson and Helmut Wenzl, who have helped by reading much of the manuscript and have picked up countless errors, we say thank you. Undoubtedly, through no fault of theirs, many, both serious and not so serious, remain.

Madison, Wisconsin   Y. AUSTIN CHANG
Salford, United Kingdom   W. ALAN OATES
October 2009

Quantities, Units, and Nomenclature

QUANTITIES AND UNITS

We have tried, throughout, to stick with the recommendations of the IUPAC (International Union of Pure and Applied Chemistry) and IUPAP (International Union of Pure and Applied Physics) on quantities, units, and symbols and also for the labeling of graph axes and table headings. See the following references:

1. Mills et al., *Quantities, Units and Symbols in Physical Chemistry*, 2nd ed., Blackwell, Oxford, 1993
2. IUPAC Report, Notation for states and processes, significance of the word "standard" in chemical thermodynamics, and commonly tabulated forms of thermodynamic functions, *J. Chem. Thermo.*, **14** (1982), 805–815.

Physical quantity = numerical value × unit

SI Units (SI = Systéme International) are used.

Primary Quantity Name	Symbol	Unit	Symbol
Length	l	Meter	m
Mass	M	Kilogram	kg
Time	t	Second	s
Electric current	I	Ampere	A
Thermodynamic temperature	T	Kelvin	K
Amount of substance	n	Mole	mol
Luminous intensity	I_v	Candela	cd

The amount of substance is of special importance to us. Previously referred to as the *number of moles*, this practice should be abandoned since it is wrong to confuse the name of a physical quantity with the name of a unit. The amount of substance is proportional to the number of specified elementary entities of that substance, the proportionality factor being the same for all substances: The reciprocal of this proportionality constant is the *Avogadro constant*. An acceptable abbreviation for amount of substance is the single word *amount*.

The elementary entities may be chosen as convenient. The concept of formula unit is important. One mole of Fe_2O_3 contains 5 mol of atoms, 1 mol of AlNi contains 2 mol of atoms, and 1 mol of $Al_{0.5}Ni_{0.5}$ contains 1 mol of atoms.

Derived Quantity Name	Unit	Symbol[a]
Force	Newton	$N\ (m\ kg^{-2})$
Pressure (or stress)	Pascal	$Pa\ (N\ m^{-2})$
Energy (or work or quantity of heat)	Joule	$J\ (N{\cdot}m)$
Surface tension	Newton per meter	$N\ m^{-1}$
Heat capacity (or entropy)	Joule per kelvin	$J\ K^{-1}$
Specific heat capacity, specific entropy	Joule per kilogram kelvin	$J\ kg^{-1}\ K^{-1}$
Specific energy	Joule per kilogram	$J\ kg^{-1}$
Molar energy	Joule per mole	$J\ mol^{-1}$
Molar heat capacity (or entropy)	Joule per mole kelvin	$J\ mol^{-1}\ K^{-1}$

[a]Symbols in parentheses refer to primary units.

NOMENCLATURE

Some IUPAC recommendations which are particularly relevant to these notes are as follows:

1. Number of entities (atoms, molecules, formula units) N
2. Amount n
3. Mass M
4. A specific quantity refers to per-unit mass, a molar quantity to per-unit amount of substance. We use lowercase for specific and subscripted with m for molar. In the case of volume, for example,

$$v = \frac{V}{M} \qquad V_m = \frac{V}{n}$$

5. Avogadro's constant L or N_A
6. Relative atomic mass (atomic weight) A_r; relative molecular mass (molecular weight) M_r. The terms atomic and molecular weights are obsolete.
7. Mass fraction w; mole fraction x; sublattice mole fraction $y_i^{(j)}$ for species i on sublattice j
8. Total pressure p; partial pressure p_i
9. Partial molar quantities are written as ΔG_{Fe} and not as $\Delta \overline{G}_{Fe}$. The subscript is sufficient to denote that it is a partial quantity without any additional bar over the symbol.
10. Accepted notations for state of aggregation are g for gas, l for liquid, s for solid. Further distinctions like cr for crystalline, vit for vitreous, and so on, are acceptable.

11. Methods for denoting processes: Any of the following may be used:

$$\Delta_l^g H^\circ = \Delta_{vap} H^\circ = H^\circ(g) - H^\circ(l)$$

Accepted abbreviations for vaporization, sublimation, melting are vap, sub, and fus, respectively.

The symbol for mixing (formation from pure components of the same structure as the solution) is mix as a subscript, for example, the molar enthalpy of mixing is written as

$$\Delta_{mix} H_m (500 \text{ K})$$

The symbol for formation (formation from pure components in their stable state at the temperature of interest) is f and so the standard molar enthalpy of formation is written as

$$\Delta_f H_m^\circ (298.15 \text{ K})$$

The symbol for reaction is r and so the standard molar Gibbs energy of reaction is written as

$$\Delta_r G_m^\circ (1000 \text{ K})$$

A process under fixed conditions is written with the function followed by | with the conditions as a subscript, for example, $dG|_{T,P}$

12. The notation for pure substance when not in its standard state is *, for example H^*.
13. The ideal state should be superscripted, for example, $\Delta_{mix} H_m^{id}$.
14. Note that unit names are always in lowercase, even if the symbols are in uppercase. This is the case even where the quantity has been named after an eminent scientist. Thus we write zero kelvin or 0 K but not zero Kelvin.
15. There are no dots in unit abbreviations, for example, $r = 10$ cm (not $r = 10$ cm.).
16. There are no dots between unit symbols, for example, J mol^{-1} K^{-1}
17. It is often advantageous to express many molar thermodynamic quantities in dimensionless form, for example, G/RT or S/R.
18. It is occasionally advantageous to use more than one subscript separated by commas. Subscripts to subscripts should be avoided.
19. Graph and table labeling should be like T/K and not T (K). This removes confusion with some of the more complicated quantities.

1 Review of Fundamentals

The following brief notes cover some of the more important points which students have met in previous courses on thermodynamics.

A principal objective of thermodynamics is to provide relations between certain equilibrium properties of matter. These relations lead to predictions about unmeasured properties. Thus, redundant measurements can be avoided, as the following sketch illustrates.

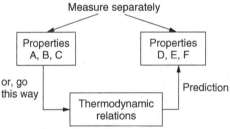

These thermodynamic relations sometimes connect quantities which might not appear to be related at first glance. An important example in regard to the subject of this book is illustrated in the following sketch:

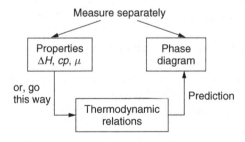

It is not immediately obvious that the phase diagram shown in the second sketch, traditionally obtained from thermal analysis measurements, can be calculated, in principle, from appropriate thermochemical measurements: Thermodynamics is concerned with the *macroscopic* properties of substances and systems *at equilibrium* (the definition of equilibrium is given later). Statistical mechanics is concerned with interpreting the equilibrium macroscopic properties in terms of *microscopic* properties, that is, in terms of atoms, electrons, bonds

Materials Thermodynamics. By Y. Austin Chang and W. Alan Oates
Copyright 2010 John Wiley & Sons, Inc.

between atoms, and so on. Specification of a microscopic state requires $\approx 10^{23}$ independent variables, usually called degrees of freedom in thermodynamics, whereas specification of a macroscopic state requires only a few independent variables (two in the case of a pure substance undergoing $p-V$ work only). The reason for this enormous reduction in the number of independent variables is that the macroscopic properties are determined by the time average of the many possible microscopic states.

Most of this course is concerned with macroscopic thermodynamics, but we will also cover some elementary aspects of statistical mechanics.

Historical Perspective   Newton (1687) quantified the concepts of force and physical work (= force × distance) but never mentioned energy. This concept came much later from Thomas Young (1807) and Lord Kelvin (1851), the latter appreciating that energy was the primary principle of physics. The science of mechanics is concerned with applying the conservation of energy to physical work problems.

Energy is the capacity to do work, potential energy being the form by virtue of position and kinetic energy being the form by virtue of motion.

There is no mention of heat in mechanics. The early calorific theory of heat had to be discarded following the experiments of Count Rumford (1798) and Joule (ca. 1850), who showed the equivalence of work transfer and heat transfer; that is, they are simply different forms of energy transfer. Work is energy transferred such that it can, in principle, be used to raise a weight, while heat is energy transferred as a result of a temperature difference. Atomistically, in work transfer, the atoms move in a uniform fashion while in heat transfer the atoms are moving in a disorganized fashion.

The equivalence of work transfer and heat transfer led to a broadening of the meaning of the conservation of energy and this became the first law in the new science of thermodynamics.

Later developments came from Carnot, Lord Kelvin, Clausius, and Boltzmann with the realization that there are some limitations in the heat transfer–work transfer process. This led to the idea of the *quality* of energy and the introduction of a new quantity, entropy. The limitations on different processes could be understood in terms of whether there is an overall increase in the thermal and/or positional disorder.

1.1 SYSTEMS, SURROUNDINGS, AND WORK

In thermodynamics we consider the system and its surroundings. It is up to the thermodynamicist to define the system and the surroundings. The two might be

  (i) isolated from one another, an *isolated* system;
 (ii) in mechanical contact only, an *adiabatic* system;
(iii) in mechanical and thermal contact, a *closed* system; or
(iv) also able to exchange matter, an *open* system.

By mechanical contact we mean that work can be exchanged between the system and the surroundings. As is illustrated schematically in Figure 1.1, *work is always measured in the surroundings* and not in the system. For the moment, we consider only mechanical work; other types of work are consider later in Section 1.5.2.

Convention   Work done *by* the system on the surroundings is taken as *positive*. Mechanical work is defined as the product of a generalized force f and its *conjugate* displacement variable dX:

$$\delta w = f_{\text{surr}}\, dX \tag{1.1}$$

The subscript surr refers to the surroundings.

Note that we write δw and not dw because work exchanged between system and surroundings is a *path-dependent* quantity. Paths may be drawn in state space (the space spanned by the chosen independent variables) with many different paths being possible in the joining of two points. Consider, for example, the two paths in going from A to B in Figure 1.2. Clearly, if we go along the path ACB the work done ($\int p\, dV$) is different from when we go along the path ADB. No work is done along the paths AC and BD (the volume is constant) with these transitions being made possible by heat transfer. This path dependence demonstrates that work is not a state function. which is defined as one which is path independent when considering movement between two points in state space.

Any state function Y, being path independent, is zero when a system is put through a cyclic path or loop, that is, for a state function,

$$\oint dY = 0$$

Especially important is the path where equilibrium is maintained, between system and surroundings, at all points as the path is traversed—a *quasi-static* or *reversible* path. Although impossible to achieve in practice, since we would have to go infinitely slowly, it is a very useful concept. When $f_{\text{surr}} = f$, the latter being

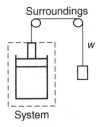

Figure 1.1 Work is measured in the surroundings and not in the system. On our convention, work done *by the system* is taken as positive.

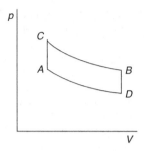

Figure 1.2   Work is not a state function; it depends on the path taken. The work done on going from A to B via ACB is different from that along the path ADB.

the value in the system, we can write

$$\delta w_{\mathrm{rev}} = f \, dX \tag{1.2}$$

An *equation of state* (EOS) is a relation between conjugate (defined later) work variables for a body in equilibrium. Some well-known examples of approximate EOS are

$$\sigma = k\epsilon \quad \text{(Hooke's law)}$$
$$pV = nRT \quad \text{(perfect gas law)}$$
$$p = -\frac{B_T}{V_0} \Delta V \quad \text{(solid compression)}$$

Given an EOS, we can then evaluate $w_{\mathrm{rev}} = \int \delta w_{\mathrm{rev}}$ along quasi-static or reversible paths. For the above EOS examples

$$w_{\mathrm{rev}} = \int \sigma \, d\epsilon = \int k\epsilon \, d\epsilon = \frac{1}{2} k\epsilon^2$$
$$(w_{\mathrm{rev}})_T = -\int p \, dV = nRT \, \log_e \left(\frac{V_2}{V_1} \right)$$
$$w_{\mathrm{rev}} = -\int p \, dV = \frac{1}{2} \frac{B_T}{V_0} (\Delta V)^2$$

1.2   THERMODYNAMIC PROPERTIES

Thermodynamic properties may be classified into being either *extensive* or *intensive*.

1. The meaning of extensive is clear. If M is mass and k a constant, then, in the case of volume, for example,

$$V(kM) = kV(M) \tag{1.3}$$

Mathematically, extensive properties like V are said to be homogenous functions of the first degree.

2. Intensive properties can be divided into two types and it is important to distinguish between the two:

   (a) *Field*: T, p, μ (much more later about this function)

   (b) *Density*: $V_m = V/N_{total}$, $H_m = H/N_{total}$, and mole fraction $x_i = N_i/N_{total}$

There is an important distinction between these two kinds of intensive variables in that a field variable takes on identical values in any coexisting phases at equilibrium, a density variable does not.

1.3   THE LAWS OF THERMODYNAMICS

The laws of thermodynamics can be introduced historically via experimental observations and many equivalent statements are possible. Alternatively, they may be stated as postulates, axiomatic statements, or assumptions based on experience. In this approach, the existence of some new state functions (bulk properties) is postulated with a recipe given for how to measure each of them. This latter approach is adopted here.

(a) *Zeroth Law*   Thermodynamic temperature T is a state function.

   Recipe   The thermodynamic temperature is equal to the ideal gas temperature, pV_m/R. It is possible, therefore, to define T in terms of mechanical ideas only, with no mention of heat. Note, however, that the thermodynamic temperature is selected as a primary quantity in the SI system.

(b) *First Law*   The internal energy U is a state function.

   Recipe   If we proceed along an adiabatic path in state space, then

$$dU = -\delta w_{adiabatic} \qquad (1.4)$$

The negative sign here arises since, if work is done by the system, its energy is lowered. Note that only changes in U can be measured. This applies to all energy-based extensive thermodynamic quantities.

The first law leads to the definition of heat. Heat should only be referred to as an energy *transfer* and not as an energy or heat *content*; that is, heat is not a noun, heat flow is a process.

For a nonadiabatic process, the change in U is no longer given by the work done on the system. The missing contribution defines the heat transferred:

$$\boxed{dU = \delta q - \delta w} \quad \text{(first law)} \qquad (1.5)$$

Just as δw is path dependent, δq is also path dependent; that is, q is not a state function, whereas U is.

Equation (1.5) is the differential form of the conservation of energy or first law for any system. Note that there is no specific mention of p–V work in this statement. It is generally valid.

Convention   Heat flow *into* the system is taken to be *positive* (Fig. 1.3). Since both work and heat flow are measured in the surroundings, where the field variables are taken to be constant, any changes in state in the surroundings are always considered to be made quasi-statically.

(c) *Second Law*   while the first law is concerned with the conservation of energy, the second law is concerned with how energy is spread. Any spontaneous process occurs in a way so as to *maximize the spread of energy between accessible states of the system and its surroundings*. Entropy is the property which is the measure of this spread.

The second law is usually stated in two parts:

1. Entropy S of the system is a state function.
   Recipe   If the state of a system is changed reversibly by heat flow, then the entropy change is given by

$$\boxed{dS = \frac{\delta q_{\text{rev}}}{T}} \quad \text{(second law, part 1)} \qquad (1.6)$$

2. In a spontaneous process, entropy in the system plus surroundings, sometimes called the universe, is created (energy is spread).

   The total entropy change of the system plus surroundings is then given by

$$dS_{\text{univ}} = dS + dS_{\text{surr}} \geq 0 \qquad (1.7)$$

In an isolated system there is no external creation of entropy so that

$$dS_{\text{surr}} = 0 \quad \text{and} \quad dS \geq 0$$

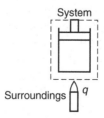

Figure 1.3   Heat flow is measured in the surroundings. In our convention, heat flow *into the system* is taken as positive.

(d) *Third Law* The third law is not really a law of macroscopic thermodynamics since its formulation requires some microscopic information. The
word law is too strong for a rule which is known to have exceptions. In a
form due originally to Planck, it can be stated as:

For any pure substance in a stable, perfectly crystalline form at 0 K, S can
be taken to be zero.

Note that:

(i) This is not to say that the entropy has an absolute value of zero.
Given time enough, all systems would undergo intranuclear and
isotopic changes. These are so slow, however, that they may be
considered to make a time-independent contribution; that is, they
contribute an additive constant to the entropy so that it is satisfactory to take this as zero.

(ii) Mixtures are specifically excluded from the defining statement.
Thus glasses, solid solutions, and asymmetric molecules may have
residual entropies at 0 K.

(iii) Pressure is not mentioned in the defining statement. This is because
$dS/dp = -dV/dT$ and the thermal expansivity $\alpha = (1/V)\, dV/dT$
is also zero at 0 K [see (1.28) for the relation between dS/dp and
β].

This wording of the third law means that the entropy of every pure crystalline
substance (element or compound) in its lowest energy state is taken to be zero
at 0 K. This wording does not preclude, for example, that $S(C(\text{diamond})) =
S(C(\text{graphite})) = 0$ at $T = 0$ K. Although there is an energy difference between
these two allotropes, the lower energy graphite states are not accessible to diamond at low temperature: Only excitations to other diamond states are possible.
This is why both C(graphite) and C(diamond) can be given zero entropies at 0 K.

It is clear that microscopic (crystallographic) information about the substance
is needed in order to be sure that the substance is in its lowest energy state.
Specifying the composition of the substance alone is not sufficient. This is why
the third law cannot be regarded on the same macroscopic footing as the zeroth,
first, and second laws.

Example 1.1 State Functions
The changes in H (or any other state function) when a system is put through a
cycle is given by

$$\oint dH = 0$$

Consider the cycle shown in Figure 1.4. Each stage in the cycle is carried out
at a total pressure of 1 bar. We place no restriction on the steps that take place
(they do not have to be carried out quasi-statically) as long as the system at the
start and end points of each step is in internal equilibrium:

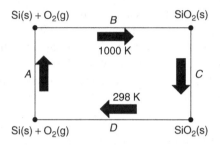

Figure 1.4   Enthalpy is a state function.

(A) $\left[H^{\circ}(1000) - H^{\circ}(298)\right]_{\text{Si(s)}} + \left[H^{\circ}(1000) - H^{\circ}(298)\right]_{O_2(g)}$

$\quad = \int_{298}^{1000} [C_p(\text{Si}) + C_p(O_2)]\, dT = 17{,}075 + 22{,}694 \text{ J}$

(B) $-\Delta_f H^{\circ}(\text{SiO}_2(s), 1000) = -857{,}493 \text{ J}$

(C) $\left[H^{\circ}(298) - H^{\circ}(1000)\right]_{\text{SiO}_2}$

$\quad = \int_{1000}^{298} C_p(\text{SiO}_2)\, dT = -43{,}611 \text{ J}$

(D) $\Delta_f H^{\circ}(\text{SiO}_2(s), 298.15) = +861{,}335 \text{ J}$

from which, for the cycle

$$\oint (A - B - C - D) = 17{,}075 + 22{,}694 - 857{,}493 - 43{,}611 + 861{,}335 = 0$$

The same procedure may be followed for the state properties $U, S, A,$ and G. For all of these state functions, $\oint dY = 0$.

1.4   THE FUNDAMENTAL EQUATION

The combined statement of the first and second laws comes by first expressing the first law for *any* process,

$$\delta q = dU + \delta w \tag{1.8}$$

and then introducing the second law for a *reversible* process, $\delta q_{\text{rev}} = T\, dS$, to obtain

$$T\, dS = dU + \delta w_{\text{rev}} \tag{1.9}$$

For a *closed system of fixed amounts of substances doing p–V work only* we can write $\delta w_{\text{rev}} = p\, dV$ so that

$$\boxed{dU = T\, dS - p\, dV} \quad (p\text{–}V \text{ work only, fixed amounts}) \tag{1.10}$$

Although we have derived this equation by considering reversible processes, it is applicable to any process as long as the initial and final states are in internal equilibrium, since it involves state functions only. It is called the fundamental equation or Gibbs's first equation and we consider its application later.

It can be seen from (1) that the natural independent variables of the state function U are S and V.

1.5   OTHER THERMODYNAMIC FUNCTIONS

For systems undergoing $p-V$ work only, we have seen that the primary functions of thermodynamics are the mechanical variables p and V together with the variables T, U, and S. For convenience, however, many other state functions are defined since it is usually not convenient for the natural variables of a system to be S and V; that is, we do not usually hold these variables constant when carrying out experiments. The introduction of new state functions enable us to change the natural variables to anything desired (in mathematical terms, we perform Legendre transformations).

The most important of these new derived functions are as follows:

1. Enthalpy H is defined as

$$H = U + pV \tag{1.11}$$

   Its usefulness comes from the fact that, at constant p,

$$dH|_p = dU + p\,dV \tag{1.12}$$

   and, if this equation is compared with

$$dU = \delta q - p\,dV \tag{1.13}$$

   then we see that

$$dH|_p = \delta q \tag{1.14}$$

   Note that there is nothing in this last equation about maintaining constant T or carrying out the process reversibly. An enthalpy change can be obtained from the measured heat flow required to bring about the change at constant p. This is the basis of *calorimetry*.

2. Heat capacities C_p and C_V are two response functions (partial derivatives of other functions):

$$C_V = \left(\frac{\partial U}{\partial T}\right)_V \qquad C_p = \left(\frac{\partial H}{\partial T}\right)_p$$

3. Helmholtz energy A is defined as $A = U - TS$. For an isothermal process $dA = dU - T\,dS$, but for a reversible process $\delta w_{\text{rev}} = -dU + T\,dS$, so

that, for a reversible, isothermal process,

$$-dA|_T = \delta w_{\text{rev}} \tag{1.15}$$

that is, in an isothermal process, the decrease in A measures the maximum work performed by the system.

4. Entropy change has already been defined: $\delta q_{\text{rev}} = T\,dS$, where δq_{rev} can be expressed in terms of the heat capacity at constant pressure. This then gives

$$dS|_p = \frac{\delta q_{\text{rev}}|_p}{T} = \frac{C_p}{T}dT \tag{1.16}$$

and, when this is integrated, advantage is taken of the third law to obtain absolute entropies:

$$S|_p(T) = \int_0^T \frac{C_p}{T}dT \tag{1.17}$$

In practice, it is more useful to do the integration in two stages:

$$S|_p(T) - S|_p(298\text{ K}) = \int_{298\text{K}}^T \frac{C_p}{T}dT \tag{1.18}$$

5. Gibbs energy G is defined as $G = U + pV - TS$. For an isothermal, isobaric process

$$dG = dU + p\,dV - T\,dS \tag{1.19}$$

For a reversible isothermal, isobaric process (combine with $\delta w_{\text{rev}} = -dU + T\,dS$),

$$-dG|_{p,T} = \delta w_{\text{rev}} - p\,dV \tag{1.20}$$

This is the total reversible work less the $p-V$ work so that, in an isothermal, isobaric process, the decrease in G measures *the maximum non−p−V work performed*. It is the most widely used derived function in materials thermodynamics. The non−p−V work of most interest to us is chemical work.

By using the definitions of the derived functions H, A, G, we can derive the other three Gibbs equations for $p-V$ work only, fixed amounts:

$$\boxed{dH = T\,dS + V\,dp} \tag{1.21}$$

$$\boxed{dA = -S\,dT - p\,dV} \tag{1.22}$$

$$\boxed{G = -S\,dT + V\,dp} \tag{1.23}$$

The natural variables of G are p, T, which are the ones usually controlled in experiments and this accounts for the importance of this particular state function.

1.5.1 Maxwell's Equations

Return to (1.10), which is the total differential of $U = U(S, V)$. We can rewrite this equation in terms of partial derivatives as follows:

$$dU = \left(\frac{\partial U}{\partial S}\right)_V dS + \left(\frac{\partial U}{\partial V}\right)_S dV \qquad (1.24)$$

If we compare (1.10) with (1.24), we see that

$$\left(\frac{\partial U}{\partial S}\right)_V = T \qquad (1.25)$$

$$\left(\frac{\partial U}{\partial V}\right)_S = -p \qquad (1.26)$$

Application of standard partial differentiation theory like this to the other Gibbs equations leads to similar relations and further relations can be obtained from the cross-derivatives, for example,

$$\left(\frac{\partial^2 U}{\partial S\, \partial V}\right)_V = \left(\frac{\partial^2 U}{\partial V\, \partial S}\right)_S \qquad (1.27)$$

which, using (1.25), gives

$$-\left(\frac{\partial p}{\partial S}\right)_V = \left(\frac{\partial T}{\partial V}\right)_S \qquad (1.28)$$

Such relations are called Maxwell's equations. Their importance lies in the fact that they can point to the recognition of redundant measurements and offer the possibility of obtaining difficult-to-measure property variations from variations in properties which are easier to measure.

All the equations in this section apply to systems performing p–V work only and are of fixed composition. We must now consider the modifications brought about by the inclusion of other types of work and the effect of changes in the amounts of substances which comprise the system.

1.5.2 Defining Other Forms of Work

With the conservation of energy as the fundamental principle, it is possible to invent other forms of thermodynamic work which can then be incorporated into the conservation-of-energy equation. By doing this, force and displacement are used in a much broader sense than they are in mechanical work. Any form of work which brings about a change in internal energy is to be considered. It may be a potential times a capacity factor or a field times a polarization.

Most notably in the present context is the invention, made by Gibbs, of chemical work. Chemical work can, of course, be considered as originating in the potential and kinetic energies of the atoms and electrons, but Gibbs realized that it is more useful to regard it as a separate form of work. In doing so, he introduced a most important new state function, the *chemical potential*, and also extended the fundamental equations to incorporate this new form of work.

The fundamental equations previously given apply to *closed* systems, that is, of fixed amounts of substance. They can be extended to include varying amounts, either for the case of a closed system, in which the amounts of substances are varying due to chemical reactions occurring within the system, or to *open* systems, where substances are being exchanged with the surroundings and in which reactions may or may not be occurring. In both cases chemical work is involved; that is, changes in internal energy are occurring.

If the amounts of substances can vary in a system, then, clearly, the state functions will depend on the n_i. In the case of U, for example, we now have $U = U(S, V, n_1, n_2 \ldots)$. Equation (1.24) will be modified to

$$dU = \left(\frac{\partial U}{\partial S}\right)_{V,n_j} dS + \left(\frac{\partial U}{\partial V}\right)_{S,n_j} dV + \sum_i \left(\frac{\partial U}{\partial n_i}\right)_{S,V,n_j} dn_i \qquad (1.29)$$

where n_j means all the others except i.

In order to be able to write this in a manner similar to (1.10), we need a symbol for the partial derivative of U with respect to n_i. The usual symbol is μ_i and its name is the *chemical potential* (p–V and chemical work only):

$$dU = T\,dS - p\,dV + \sum_i \mu_i\,dn_i \qquad (1.30)$$

The other Gibbs equations may be modified in a similar fashion (p–V and chemical work only):

$$dH = T\,dS + V\,dp + \sum_i \mu_i\,dn_i \qquad (1.31)$$

$$dA = -S\,dT - p\,dV + \sum_i \mu_i\,dn_i \qquad (1.32)$$

$$dG = -S\,dT + V\,dp + \sum_i \mu_i\,dn_i \qquad (1.33)$$

Note that the definition of μ_i varies depending on which function is being used:

$$\mu_i = \left(\frac{\partial U}{\partial n_i}\right)_{S,V,n_j} = \left(\frac{\partial H}{\partial n_i}\right)_{S,p,n_j} = \left(\frac{\partial A}{\partial n_i}\right)_{T,V,n_j} = \left(\frac{\partial G}{\partial n_i}\right)_{T,p,n_j} \qquad (1.34)$$

This extension of thermodynamics from a study of heat engines to its application to phase and chemical equilibrium by Gibbs represents one of the greatest achievements in nineteenth-century science. Recall that, in the application of thermodynamics to heat engines, the nature of the fluid of the engine is unimportant, but in introducing chemical work, the nature of the material constituting the system becomes all important.

Chemical work is not the only possible form of work which might have to be considered. Some examples of different types of thermodynamic work are given in the accompanying table. In order to emphasize the fact that (1.10) is a restricted form of the fundamental equation, let us write down a more complete statement which takes into account some other possible types of work:

$$dU = T \, dS - p \, dV + \sum_{i=1}^{N} \mu_i \, dn_i + \phi \, dm$$

$$+ \sum_{i=1}^{N} \psi_i \, dQ_i + \sigma \, d\epsilon + \gamma \, dA_s + \cdots \qquad (1.35)$$

In writing the equation in this form it is assumed that the various types of work are independent of one another. This may not always be the case; often, two or more work terms are *coupled*.

A complete new set of Maxwell's equations can also be obtained from the fundamental equations for systems undergoing these various forms of work.

Type of Work	Field Variable	Extensive Variable	Differential Work in dU
Mechanical			
Pressure–volume	$-p$	V	$-p \, dV$
Elastic	τ_{ij}	$V \eta_{ij}$	$V \sum \tau_{ij} \, d\eta_{ij}$
Gravitational	$\phi = gh$	$m = \sum M_i n_i$	$\phi \, dm = \sum gh M_i \, dn_i$
Surface	γ	A_s	$\gamma \, dA_s$
Electromagnetic			
Charge transfer	ψ_i	Q_i	$\psi_i \, dQ_i$
Electric polarization	$\mathbf{E}$	p	$\mathbf{E} \cdot dp_i$
Magnetic polarization	$\mathbf{B}$	m	$\mathbf{B} \cdot dm$
Chemical			
No reactions	μ_i	n_i species	$\sum_i \mu_i \, dn_i$
With reactions	μ_i	ξ extent of reaction	$\sum_i \nu_i \mu_i \, d\xi$

1.6   EQUILIBRIUM STATE

A precise definition of a system in equilibrium is not straightforward. To define a system as being in equilibrium when its properties are not changing with time is unacceptable—the state which involves a steady flow of heat or matter through a system is a time-independent state but systems in which these processes are occurring are not in equilibrium; there are field gradients. We need a better definition and one is discussed below.

The second law, part 2, states that $dS_{\text{univ}} \geq 0$, with the inequality referring to spontaneous processes and the equality to reversible processes, the latter corresponding with the system being in equilibrium.

If we consider an isolated system (no work or heat flow and, therefore, constant U and V), then $dS_{\text{surr}} = 0$ so that

$$dS|_{U,V} = dS_{\text{univ}} \geq 0 \qquad (1.36)$$

In other words, for an isolated system, S reaches a maximum at the equilibrium state, making this state function the appropriate *thermodynamic potential* for isolated systems. The important point here is that, under certain constraints, we have replaced a property of the universe (system + surroundings) by a property of the system alone.

Of more practical interest is to obtain the appropriate thermodynamic potential for constant p and T conditions and the nature of the extrema conditions. We can do this as follows:

$$G = U + pV - TS \qquad (1.37)$$

$$dG = dU + p\,dV + V\,dp - T\,dS - S\,dT \qquad (1.38)$$

$$= \delta q - p_{\text{surr}}\,dV + p\,dV + V\,dp - T\,dS - S\,dT \qquad (1.39)$$

and at constant p and T where $p_{\text{surr}} = p$ we have

$$dG|_{p,T} = -T\,dS_{\text{surr}} - T\,dS$$

$$= -T\,dS_{\text{univ}} \leq 0 \qquad (1.40)$$

From the general statement of the second law, dS_{univ} is a maximum at equilibrium, it follows from (1.40) that *the appropriate thermodynamic potential for conditions of constant p and T is the Gibbs energy and G evolves to a minimum at equilibrium*. Since these conditions are the most frequently met, the Gibbs energy is usually the most important thermodynamic potential of interest. Note that, again, a property of the universe has been replaced by a property of the system alone.

For small excursions from an equilibrium state, we can expand any function for G as a Taylor series in the state space variables. As illustrated in Figure 1.5, which shows G as a function of only two state space variables, the extrema in

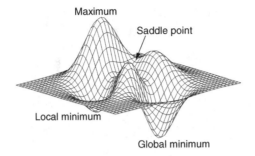

Figure 1.5 Local and global equilibrium, drawn using Matlab®.

multidimensional space can be maxima, minima, or saddle points (a maximum in some directions, a minimum in others). The required conditions for the extremum to be a minimum when there are two such variables, x and y, can be written as

$$\frac{\partial^2 G}{\partial x^2} \quad \text{and} \quad \frac{\partial^2 G}{\partial y^2} > 0 \tag{1.41}$$

$$\frac{\partial^2 G}{\partial x^2} \frac{\partial^2 G}{\partial y^2} > \left(\frac{\partial^2 G}{\partial x\, \partial y} \right)^2 \tag{1.42}$$

Failure of the condition given in (1.40) implies a saddle point.

These conditions only apply, however, for small excursions from the equilibrium point. As shown in Figure 1.5, it is possible to have a local minimum which fulfils the above conditions, but it is not the global minimum which we seek in our thermodynamic calculation.

For both the local and global minima, a small fluctuation from the equilibrium point will result in $dG|_{p,T} > 0$ and the system will wish to return to its equilibrium point. In both cases also the field variables (T, p, μ_A) are constant throughout the system. This means that we can apply the equations of thermodynamics equally well to the metastable local equilibrium and the stable global equilibrium situations if we ensure that there are no large-scale fluctuations which will take us from the local to the global equilibrium.

The global equilibrium, that is, *the true equilibrium state, is when* $\Delta G|_{p,T} > 0$ *for any excursions from that state*, providing the start and end states are maintained in internal equilibrium by the imposition of extra constraints. It is this definition of the equilibrium state which is mainly used throughout these chapters, but, as has been indicated previously, other thermodynamic potentials fulfil the same role as G for other conditions.

EXERCISES

1.1 Starting from Al(s) and O_2(g) at 298 K and 1 bar, use the data given below and the cycle illustrated in Figure 1.6 to confirm that $\oint dS = 0$:

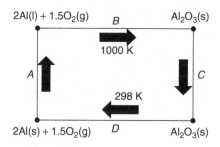

Figure 1.6   Cycle to be considered.

$$T_{\text{fus}}(\text{Al}) = 933 \text{ K} \qquad \Delta_{\text{fus}} H_m = 10460 \text{ J mol}^{-1}$$

$$C_p = a + bT + cT^2 + dT^{-2}$$

	a	$b/10^3$	$c/10^6$	$d/10^{-5}$	$S^{\circ}(298)/\text{J K}^{-1} \text{ mol}^{-1}$
Al(s)	20.7	12.4	0	0	28.3
Al(l)	31.8	0	0	0	0
O_2(g)	30	4.184	0	−1.67	205.0
Al_2O_3(s)	106.6	17.78	0	−28.53	51.0

1.2   Starting with $CaCO_3$(s) at 298 K and 1 bar:

(a) Calculate the heat transferred in producing 1 mol of CaO(s) at 1200 K and 1 mol of CO_2(g) at 500 K.

(b) Calculate the standard entropy change for this process.

1.3   (a) Write down the full equations required for evaluation of the standard enthalpy and Gibbs energy for the reaction equation

$$CaCO_3(\text{s}) = CaO(\text{s}) + CO_2(\text{g})$$

(b) Determine the temperature at which $\Delta G^{\circ} = 0$ for this reaction equation.

(c) Calculate the enthalpy of reaction at the temperature for which the equilibrium pressure of CO_2 is 1 bar.

Substance	$\Delta_f H^{\circ}(298\text{K})/$ kJ mol^{-1}	$S^{\circ}(1000K)/$ J K^{-1} mol^{-1}	$C_p = a + bT + cT^2 + dT^{-2}/\text{J K}^{-1} \text{ mol}^{-1}$			
			a	$b \times 10^3$	$c \times 10^5$	$d \times 10^{-5}$
CaO(s)	−634.92	96.96	57.75	−107.79	0.53	−11.51
CO_2(g)	−393.51	269.19	44.14	9.04	0	−8.54
$CaCO_3$(s)	−1206.60	220.21	99.55	27.14	0	−21.48

1.4 Given the following data:
Note that $\langle C_p^{\circ} \rangle$ refers to the average values of C_p° over the range 298–1000 K. Calculate:

(a) The standard entropy of oxidation of Si(s) to SiO_2(s) at 1000 K.

(b) The same using the values of $\langle C_p^{\circ} \rangle$. Compare with the result from (a).

Substance	S°(s, 298 K)/ J K^{-1} mol^{-1}	$S^{\circ}_{1000} - S^{\circ}_{298}$)/ J K^{-1} mol^{-1}	$\langle C_p^{\circ} \rangle$/ J K^{-1} mol^{-1}
Si(s)	18.81	28.69	23.23
O_2(g)	209.15	38.43	32.10
SiO_2(s)	27.78	72.15	56.02

(c) $\Delta_f H^{\circ}(SiO_2, 1000\,K)$ using the value of $\Delta_f H^{\circ}(SiO_2, 298\,K)$ in the text.

1.5 Derive (1.21), (1.22), and (1.23) from (1.10) and the definitions of the functions H, A and G.

1.6 (a) If the entropy of transition of a pure substance A, $\Delta_{\alpha}^{\beta} S^{\circ}(A)$, at constant p is constant, show that the corresponding enthalpy change, $\Delta_{\alpha}^{\beta} H^{\circ}(A)$, is also constant.

(b) If the phase transition of a pure substance is a function of T and p and the value of $\Delta_{\alpha}^{\beta} S^{\circ}(A)$ is independent of the change in conditions, show that the value of $\Delta_{\alpha}^{\beta} H^{\circ}(A)$ is no longer constant, as was the case for the constraint of constant p. (*Hint*: Use the Maxwell relationships.)

2 Thermodynamics of Unary Systems

A unary, or one-component, system refers to a pure substance of fixed composition. Importantly, this includes molecules and compounds as well as the pure elements. For molecules and compounds, their formation properties are often the most important. Formation properties are defined as the properties of the compound relative to those of the elements in their most stable form at the temperature and pressure of interest.

In this chapter, only $p-V$ work is considered. Chemical work, surface work, and so on, are ignored.

2.1 STANDARD STATE PROPERTIES

Standard states, by definition, always refer to a pressure of 1 bar. Reference states, which are also used in thermodynamics, are not necessarily so constrained. We can also speak of reference states for standard state properties, for example, a reference state at 298.15 K for a standard state property at 1 bar.

For the moment we concentrate on properties at 1 bar, that is, on standard state properties.

Collating all the available experimental data for pure substances and arriving at recommended values for thermochemical properties comprise a highly skilled and major exercise. It has been done by large organizations, for example, the National Institute of Standards and Technology (NIST), the National Aeronautics and Space Administration (NASA), and the U.S. Bureau of Mines. Before computers, thermochemical data were usually presented in the form of tables; occasionally they were presented as analytical approximations. With the advent of computers, however, analytical representations have almost totally supplanted the use of tables. The storage of the coefficients in analytical expressions is much more convenient for computer usage than their presentation in tabular form.

A quantity of primary interest is the standard Gibbs energy of formation of a compound, which is related to the standard enthalpy and entropy of formation by

$$\Delta_f G^\circ(T) = \Delta_f H^\circ(T) - T \Delta_f S^\circ(T) \qquad (2.1)$$

Materials Thermodynamics. By Y. Austin Chang and W. Alan Oates
Copyright 2010 John Wiley & Sons, Inc.

From values of $\Delta_f G^\circ(T)$ for the participating substances, we can obtain standard Gibbs reaction energies, $\Delta_r G^\circ(T)$, for any reaction:

$$\Delta_r G^\circ(T) = \sum_i v_i \Delta_f G^\circ(i, T) \tag{2.2}$$

Here v_i is the stoichiometric coefficient, which is positive for products and negative for reactants in a chemical reaction, that is, for a formation reaction such as

$$Pb + \tfrac{1}{2}O_2(g) = PbO$$

$$v_{PbO} = 1 \qquad v_{Pb} = -1 \qquad v_{O_2} = -\tfrac{1}{2}$$

Standard Heat Capacities   As will be discussed in Chapter 3, the standard heat capacity of a substance varies with temperature in a rather complex manner at low temperatures. Above room temperature, however, and in the absence of any phase transformations, its variation with T can be represented to a sufficiently high accuracy by a polynomial. The approximation usually used is of the form

$$C_p^\circ(T) = a + bT + cT^2 + dT^{-2} \tag{2.3}$$

Note, however, that some data compilations use slightly different analytical expressions.

Standard Entropies of Formation   In the absence of any phase transformations, the standard entropy for a pure substance at any temperature T may be referred to its value at 298.15 K using

$$S^\circ(T) - S^\circ(298.15) = \int_{298.15}^T \frac{C_p^\circ}{T}\, dT \tag{2.4}$$

If (2.3) is used for the standard heat capacity of the substance, then we see that this standard entropy difference can be evaluated from the same stored polynomial coefficients:

$$S^\circ(T) - S^\circ(298.15) = a \log_e\left(\frac{T}{298.15}\right) + b(T - 298.15)$$
$$+ \tfrac{1}{2}c(T^2 - 298.15^2) - \tfrac{1}{2}d(T^{-2} - 298.15^{-2}) \tag{2.5}$$

When a phase transition occurs in the substance, it is necessary to allow for this in the representation. Different allotropes require different C_p° representations and there is the transformation entropy between the allotropes to be considered. If the substance undergoes a transition from α to β at temperature T_{Tr},

$$S^\circ(T) - S^\circ(298.15) = \int_{298.15}^{T_{Tr}} \frac{C_p^\circ(\alpha)}{T}\, dT + \Delta_\alpha^\beta S^\circ + \int_{T_{Tr}}^T \frac{C_p^\circ(\beta)}{T}\, dT \tag{2.6}$$

If the substance undergoes a magnetic transition, say from ferromagnetic to paramagnetic, the heat capacity variation is more complex and the simple polynomial representation has to be augmented with another representation for the magnetic contribution.

The Planck postulate (S for any pure substance in a stable perfect crystalline form at 0 K is taken to be zero) is taken advantage of in obtaining $S^\circ(298.15)$.

Although C_p° below room temperature is a complex function of temperature which cannot be represented by a simple polynomial like (2.3), once it has been measured, the standard entropy of a crystalline substance at 298.15 K can be obtained from

$$S^\circ(298.15) = \int_0^{298.15} \frac{C_p^\circ(T)}{T} \, dT \qquad (2.7)$$

This means that the computer storage of $S^\circ(298.15 \text{ K})$ plus the coefficients in the standard heat capacity equation (2.3) is sufficient for evaluating the standard entropies of pure substances in a particular structural form at any temperature above room temperature.

For a pure compound substance it is the formation property which is of major interest. In view of the Planck postulate we can write

$$\Delta_f S^\circ(298.15) = \sum_i v_i S^\circ(i, 298.15) \qquad (2.8)$$

where the summation is over all participating substances in the formation reaction equation. Equation (2.8) can then be used in the calculation of the high-temperature standard entropy of formation of the compound:

$$\Delta_f S^\circ(T) = \Delta_f S^\circ(298.15) + \sum_i v_i \left[S^\circ(i, T) - S^\circ(i, 298.15) \right] \qquad (2.9)$$

where (2.5) is used for $S^\circ(i, T) - S^\circ(i, 298.15)$.

If phase transformations occur in either products or reactants, (2.9) has to be modified in a manner similar to that given in (2.6).

Standard Enthalpies of Formation   The difference in standard enthalpy for a substance between two temperatures is given by

$$H^\circ(T_2) - H^\circ(T_1) = \int_{T_1}^{T_2} C_p \, dT \qquad (2.10)$$

Whereas standard entropies have a natural reference point by taking advantage of the Planck postulate, there is no similar natural reference point for enthalpies.

In the following we concentrate on the two most commonly used choices of reference state. The first one is used in most tabular presentations while the second one is favored in computer databases:

1. *Standard Substance Reference (SSR) State*   The enthalpies of all *substances* in their most stable form at 1 bar and 298.15 K are selected to be zero.

Table 2.1 for copper, taken from the NIST-JANAF tables, illustrates this selection. It can be seen that $H°(298.15\ K) = 0$. It also has this value for compound substances.

Using the SSR state, the enthalpy of formation of a pure compound can be obtained from

$$\Delta_f H°(T) = \Delta_f H°(298.15) + \sum_i \nu_i \left[H°(i, T) - H°(i, 298.15) \right] \qquad (2.11)$$

There is no equation analogous to (2.8) for formation enthalpies at 298.15 K. These have to be determined by experiment or calculation.

If (2.3) is used for the heat capacity, the bracketed terms on the right-hand side of (2.11) are given by

$$H°(i, T) - H°(i, 298.15\ K) = a(T - 298.15) + \tfrac{1}{2}b(T^2 - 298.15^2)$$
$$+ \tfrac{1}{3}c(T^3 - 298.15^3) - d(T^{-1} - 298.15^{-1})$$
$$(2.12)$$

In tabular form as shown in Table 2.1 for Cu, data are usually presented at 100 K intervals. If data at temperatures intermediate to these are required, it is necessary to interpolate. In order to assist in doing this with some accuracy, some function which is a slowly varying function of temperature is desired. Such a function is the so-called Gibbs energy function gef (a confusing term) defined as

$$\text{gef}(T) = -\frac{G°(T) - H°(298.15)}{T} \qquad (2.13)$$

As can be seen in Table 2.1 for Cu, this function has the desired property of being slowly varying with temperature and is therefore easily and accurately interpolated. Given the values of gef for all the species involved in the formation reaction, the calculation of $\Delta_f G°(T)$ is then straightforward. From (2.13),

$$T \times \text{gef}(T) = -G°(T) + H°(298.15)$$

and we see that $\Delta_f G°(T)$ can be obtained from

$$\Delta_f G°(T) = \Delta_f H°(298.15) - T \sum_i \nu_i \, \text{gef}(i, T) \qquad (2.14)$$

In order to evaluate $\Delta_f G°(T)$, all that is required is the room temperature standard enthalpy of formation and the tables can then be used to obtain $\sum_i \nu_i \, \text{gef}(i, T)$ for the formation of the substance. It should be noted that the difference, $\sum_i \nu_i \, \text{gef}(i, T)$), usually varies even more slowly with T than the gef values for the individual substances.

TABLE 2-1 Tabular Presentation of Thermodynamic Properties of Copper from NIST-JANAF Tables

Enthalpy Reference Temperature $T_s = 298.15$ K				Standard State Pressure $= 0.1$ MPa			
	J K^{-1} mol^{-1}			kJ mol^{-1}			
T/K	C_p°	S°	$\dfrac{-[G^\circ - H^\circ(T_s)]}{T}$	$H^\circ - H^\circ(T_s)$	$\Delta_f H^\circ$	$\Delta_f G^\circ$	log K_t
0	0	0	0	−5.007	0	0	0
100	16.010	10.034	53.414	−4.338	0	0	0
200	22.631	23.73	35.354	−2.325	0	0	0
250	23.782	28.915	33.563	−1.162	0	0	0
298.15	24.442	33.164	33.164	0.000	0	0	0
300	24.462	33.315	33.164	0.045	0	0	0
350	24.975	37.127	33.464	1.282	0	0	0
400	25.318	40.484	34.136	2.539	0	0	0
450	25.686	43.489	35.011	3.815	0	0	0
500	25.912	46.206	35.997	5.105	0	0	0
600	26.481	50.982	38.107	7.725	0	0	0
700	26.996	55.030	40.247	10.399	0	0	0
800	27.494	58.739	42.336	13.123	0	0	0
900	28.049	62.009	44.343	15.899	0	0	0
1000	28.662	64.994	46.261	18.733	0	0	0
1100	29.479	67.763	48.091	21.638	0	0	0
1200	30.519	70.368	49.840	24.633	0	0	0
1300	32.143	72.871	51.516	27.762	0	0	0
1358	33.353	74.300	52.459	29.660	Crystal ⟷ Liquid		
1358	32.844	83.974	52.459	42.798	Transition		
1400	32.844	84.974	53.419	44.177	0	0	0
1500	32.844	87.240	55.599	47.462	0	0	0
1600	32.844	89.360	57.644	50.746	0	0	0
1700	32.844	91.351	59.569	54.031	0	0	0
1800	32.844	93.229	61.387	57.315	0	0	0
2000	32.844	96.689	64.747	63.884	0	0	0
2200	32.844	99.819	67.795	70.453	0	0	0
2400	32.844	102.677	70.585	77.022	0	0	0
2600	32.844	105.306	73.156	83.591	0	0	0
2800	32.844	107.740	75.540	90.159	0	0	0
2843.3	32.844	108.244	76.034	91.580	FUGACITY $= 1$ bar		
2900	32.844	108.893	76.671	93.444	−300.204	5.996	−0.108
3000	32.844	110.006	77.764	96.728	−299.409	16.541	−0.288
3200	32.844	112.126	79.846	103.297	−297.971	37.556	−0.613
3400	32.844	114.117	81.804	109.866	−296.737	58.488	−0.899
3600	32.844	115.995	83.652	116.435	−295.703	79.352	−1.151
3800	32.844	117.770	85.401	123.004	−294.962	100.165	−1.377
4000	32.844	119.455	87.062	129.573	−294.199	120.938	−1.579

Note: The data for some temperatures have been removed to decrease the size of the table.

The NIST-JANAF compilation also presents tables for other choices of reference state. In one set of tables, as discussed above, the most stable form at 1 bar and 298.15 K is selected while, in another set, the enthalpies of all substances in the designated form, regardless of whether that form is the most stable or not, are selected to be zero at 1 bar and 298.15 K. For example, the properties for a liquid substance are referred to the liquid at 298.15 K in this other form of presentation. Other compilations select the enthalpy of the substance in its most stable form at 0 K to be zero and it is clear that some care is required if these different presentations of data are to be used correctly.

With the advent of the computer storage of data, tabular presentations are at the obsolescent stage.

2. *Standard Element Reference (SER) State*   The enthalpies of the *elements* in their most stable form at 1 bar and 298.15 K are selected to be zero.

In order to distinguish between the SSR and SER states for the standard enthalpy, we will use H^{SER} to refer to $H°(298.15)$ when the element reference state is being used. Here, H^{SER} is set to zero for the elements in their most stable state at 298.15 K while $H^{SER}(298.15)$ for a compound (abbreviation cpd) is equal to its enthalpy of formation at that temperature:

$$H^{SER}(\text{cpd}, 298.15) = \Delta_f H°(\text{cpd}, 298.15) \tag{2.15}$$

The standard formation enthalpy for a compound at any temperature on this reference state is then given by

$$\Delta_f H°(T) = \sum v_i H°(i, T) \tag{2.16}$$

an equation which can be compared with (2.11) based on using the substance reference state.

When phase transformations are present, the enthalpies of transition must be incorporated into (2.16).

Table 2.2 shows the way that the necessary coefficients, used in the SER, are stored in computer databases for PbO.

   (i) Line 1 in Table 2.2 gives the values for H^{SER} and $S°(298.15)$.

   (ii) Line 2 gives the heat capacity coefficients between 298.15 and 762 K. At this temperature there is a phase transformation in PbO(s).

   (iii) Line 3 gives the enthalpy change at this transition. The standard entropy change can be obtained from $\Delta_{Tr} H°/T_{Tr}°$ since $\Delta_{Tr} G° = 0$ at this temperature.

   (iv) Line 4 gives the heat capacity coefficients for the temperature range 720–1162 K. PbO melts at this temperature.

   (v) Line 5 gives the standard enthalpy of fusion. Again the standard entropy of fusion can be obtained from $\Delta_{fus} H°/T_{fus}°$.

   (vi) Line 6 gives the heat capacity coefficients for liquid PbO.

TABLE 2-2 Computer Stored Coefficients for Evaluating Standard State Properties of PbO

Line No.	T	$H°$	$S°$	a	b	c	d
1	298.15	−218,600.00	67.84				
2	298.15−762.0			40.9683	0.0203012	0.15180900 ×10^{-6}	−53460.602
3	762.0	200.0					
4	762.0−1160.0			38.853298	0.19788301 ×10^{-1}	0.67834202 ×10^{-6}	−369456.00
5	1160.0	25,580.000					
6	>1160.0			3200.0000	0	0	0

Note: SI units are used.

With the storage of these parameters, it is possible to calculate the standard state properties for PbO on the SER scale at any temperature.

Example 2.1 Using SSR and SER
It is essential to become familiar with manipulating thermodynamic data presented as both SSR and SER. We illustrate with the specific example of calculating $\Delta_f H°(T)$ and $\Delta_f S°(T)$ for the formation reaction at 1 bar:

$$Pb(T) + \tfrac{1}{2}O_2(g)(T) = PbO(s)(T) \qquad (2.17)$$

We will concentrate on a calculation for 800 K, where Pb(l) is the stable form of lead.

No matter whether the SSR or SER state is used, we must obtain the same calculated result for these properties.

The computer storage of data requires not only the coefficients given in Table 2.2 for PbO but similar tables of coefficients for Pb and O_2.

The formation properties are plotted in Figures 2.1a and 2.1b as a function of temperature. The large steps apparent in both curves occur at the melting point of Pb; that is, there is a difference in $\Delta_f H°(T)$ and $\Delta_f S°(T)$ at 600 K for the reaction equations:

$$Pb(s) + \tfrac{1}{2}O_2(g) = PbO(s) \qquad (2.18)$$

$$Pb(l) + \tfrac{1}{2}O_2(g) = PbO(s) \qquad (2.19)$$

The small steps in the curves shown in Figs. 2.1a and 2.1b occur at a solid-state transformation in PbO (line 3 in the Table 2.2).

The entropy change calculation is the same no matter whether the SSR or SER state is used since both utilize the same, 0 K, reference state. The standard

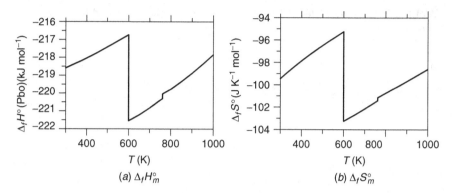

Figure 2.1 Formation properties of PbO(s) as a function of T.

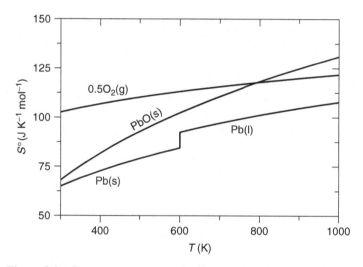

Figure 2.2 Standard molar entropies for PbO(s), Pb(l), and $0.5O_2(g)$.

entropies for the participating substances are shown in Figure 2.2. At 800 K

$$\Delta_f S^\circ(800) = S^\circ(\text{PbO}(s), 800) - S^\circ(\text{Pb}(l), 800) - 0.5 S^\circ(O_2(g), 800)$$

$$= 118.329 - 101.172 - 0.5 \times 235.819$$

$$= -100.75 \text{ JK}^{-1} \text{ mol}^{-1}$$

The value of 118.329 J K^{-1} mol^{-1} for PbO(s) has been obtained from (2.5) using the coefficients given in Table 2.2 and is also shown in Figure 2.1b.

The calculation of the standard formation enthalpy, on the other hand, differs, depending on whether the SSR or SER state is used. The standard enthalpies with

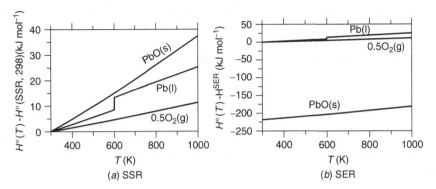

Figure 2.3 Standard molar enthalpies for PbO(s), Pb(l), and $0.5O_2$(g).

respect to the two reference states under consideration are shown in Figures 2.3a and 2.3b.

Using the SSR,

$$\Delta_f H^\circ(800) = \Delta_f H^\circ(298.15) + [H^\circ(\text{PbO(s)}, 800) - H^\circ(\text{PbO(s)}, 298.15)]$$
$$- [H^\circ(\text{Pb(l)}, 800) - H^\circ(\text{Pb(l)}, 298.15)]$$
$$- 0.5[H^\circ(O_2(\text{g}), 800)) - H^\circ(O_2(\text{g}), 298.15)]$$
$$= -218.6 + 26.143 - 19.403 - 0.5 \times 15.837$$
$$= -219.78 \text{ kJ mol}^{-1}$$

Using the H^{SER},

$$\Delta_f H^\circ(800) = H^\circ(\text{PbO(s)}, 800) - H^\circ(\text{Pb(l)}, 800) - 0.5H^\circ(O_2(\text{g}), 800)$$
$$= -192.457 - 19.403 - 0.5 \times 15.837$$
$$= -219.78 \text{ kJ mol}^{-1}$$

When the SER is used, we will usually denote the standard Gibbs energy $G^\circ(T)$ as $G(\text{SER}, T)$, where

$$G(\text{SER}, T) = [H^\circ(T) - H^{\text{SER}}] - TS^\circ(T) \tag{2.20}$$

2.2 THE EFFECT OF PRESSURE

The EOS for a pure substance is a relationship between any four thermodynamic properties of the substance, three of which are independent. Usually the EOS involves pressure p, volume V, temperature T, and amount of substance in the system, n:

$$\pi(p, V, T, n) = 0 \tag{2.21}$$

which indicates that, if any three of the four properties are fixed, the fourth is determined. More usually, the EOS is written in a form which depends only on the nature of the system and not on how much of the substance is present; hence all extensive properties are replaced by their corresponding specific values. The molar form of the above EOS is

$$\pi(p, V_m, T) = 0 \qquad (2.22)$$

If any two of these thermodynamic properties are fixed, the third is determined. Some simple EOS for gases and condensed phases which have been proposed are given below.

2.2.1 Gases

2.2.1.1 The Perfect Gas. The EOS for a perfect gas is familiar:

$$pV_m = RT \qquad (2.23)$$

We can now illustrate how knowledge of an EOS gives us the pressure variation of the thermodynamic properties at constant temperature. The fourth Gibbs equation for a pure substance is

$$dG = -S\,dT + V\,dp \quad (p\text{--}V \text{ work only}) \qquad (2.24)$$

Substituting the ideal gas EOS into the above equation, we have

$$dG = -S\,dT + \frac{nRT}{p}dp \qquad (2.25)$$

$$= -S\,dT + nRT\,d\,\log_e p \qquad (2.26)$$

At constant T, for a perfect gas only, and for 1 mol, this reduces to

$$dG_m|_T = RT\,d\,\log_e p \qquad (2.27)$$

or, in integral form,

$$\boxed{G_m^*(T, p) - G_m^\circ(T) = RT\log_e\left(\frac{p^*}{p^\circ}\right)} \qquad (2.28)$$

Here we have introduced the superscript asterisk to indicate the quantity for a pure substance at any pressure other than the standard.

2.2.1.2 Real Gases. In chemical engineering, where there is much interest in the behavior of substances in those parts of state space where the distinction between gas and liquid becomes blurred, it is necessary to have an EOS which permits the calculation of the thermodynamic properties at all temperatures and pressures. A large number of such equations has been proposed and we will meet one of these later in Chapter 4.

In the regime where the properties of a real gas do not deviate substantially from those of a perfect gas, it is common to use the concept of *fugacity*. The functional form of (2.27) is retained but the fugacity f is used to replace pressure. Fugacity has the same units as for pressure. In differential form

$$dG_m(T) = RT \, d \log_e f \quad \text{(real gas, constant } T) \tag{2.29}$$

which, in integrated form, becomes

$$G_m^*(T, p) - G_m^\circ(T) = RT \log_e \left(\frac{f^*}{f^\circ} \right) \tag{2.30}$$

Tabulations as well as graphical and analytical presentations of the fugacities of the common gases are readily available.

2.2.2 Condensed Phases

Condensed phases are much less compressible than gases and it is rare that the effect of pressure on thermodynamic properties has to be taken into account in calculations in material science and engineering (MS&E), in contrast, for example, to such calculations in applications to geochemistry and astrophysics.

The difference in the effect of pressure on solids and liquids as compared with its effect on gases is readily illustrated. At constant T for a pure substance undergoing p–V work only,

$$dG_m|_T = V_m \, dp \tag{2.31}$$

For a perfect gas at standard temperature and pressure (STP), $V_m = 22,400 \text{ cm}^3$ mol^{-1}, but for a solid or liquid V_m is much smaller, on the order of 10 cm^3 mol^{-1}. The large difference in the effect of pressure on G is due to these large differences in molar volume.

The simplest EOS which may be used to describe a solid under pressure changes at constant temperature comes from the definition of isothermal compressibility:

$$\kappa_T = -\frac{1}{V} \left(\frac{\partial V}{\partial p} \right)_T \tag{2.32}$$

If κ_T is assumed independent of pressure, then integration gives

$$\frac{V - V^\circ}{V^\circ} \approx -\kappa_T(p - p^\circ) \tag{2.33}$$

and hence, using (2.31), the Gibbs energy is

$$G_m^*(T, p) - G_m^\circ(T) \approx -\frac{V^* - V^\circ}{V^\circ \kappa_T} \tag{2.34}$$

More sophisticated expressions for the EOS for solids take into account the effect of p and T on V_m and κ_T.

Summarizing: To a good *approximation*, it is usually safe in MS&E to use

$$\boxed{G_m^*(T, p) \approx G_m^\circ(T) + RT \, \log_e \left(\frac{p^*}{p^\circ}\right)} \quad \text{(pure gases)} \tag{2.35}$$

$$\boxed{G_m^*(T, p) \approx G_m^\circ(T)} \quad \text{(pure solids, liquids)} \tag{2.36}$$

2.3   THE GIBBS–DUHEM EQUATION

Although there are the three field variables T, p, and μ to consider for a pure substance, we have previously stated that the state of a one-component, single-phase system is fully specified by only two independent variables. Thus there must be a relationship between the three field variables in a single-phase unary system. This relation is the important Gibbs–Duhem equation (we will meet it again when discussing solution phases). One way by which it may be derived is given below.

The fundamental equation for a pure substance A is

$$dU = T \, dS - p \, dV + \mu_A \, dn_A \tag{2.37}$$

Since all three independent variables in this equation are extensive, its integration gives the integral extensive quantity as follows:

$$U = TS - pV + n_A \mu_A \tag{2.38}$$

This equation can also be obtained from (2.37) by applying Euler's theorem for homogeneous functions of degree 1 (see Appendix B).

When we define $H = U + pV$ and $A = U - TS$, however, the fundamental equations are

$$dH = T \, dS + V \, dp + \mu_A \, dn_A \tag{2.39}$$

$$dA = -S \, dT - p \, dV + \mu_A \, dn_A \tag{2.40}$$

These equations now have only two extensive independent variables, the other being a field variable. The extensive integral quantities are obtained in this case by considering only the extensive variables to give

$$H = TS + n_A \mu_A \tag{2.41}$$

$$A = -pV + n_A \mu_A \tag{2.42}$$

On defining the Gibbs energy, $G = H - TS$, the resulting fundamental equation is now a function of only one extensive independent variable:

$$dG = -S\,dT + V\,dp + \mu_A\,dn_A \tag{2.43}$$

and the integral quantity is obtained by considering only this independent variable to give

$$G = n_A \mu_A \tag{2.44}$$

It is possible to go one step further by defining the new function, called the grand potential and defined as $\Omega = G - n_A \mu_A$. Using (2.44) we see that both Ω and $d\Omega$ equal 0 and there is now a new fundamental equation:

$$\boxed{0 = -S\,dT + V\,dp + n_A\,d\mu_A} \tag{2.45}$$

This is the *Gibbs–Duhem equation*, which gives the relation between the three field variables for a single-phase unary system. The Gibbs–Duhem equation confirms that the state of a single-phase unary system is fully specified by only two field variables.

2.4 EXPERIMENTAL METHODS

It is clear from the material presented earlier in this chapter that, in order to be able to calculate all the high-temperature thermodynamic properties of pure substances at 1 bar, certain experimental information is required. The measurements which are made in order to provide the necessary data rely heavily on the use of *calorimeters* which are of two types: *substance* calorimeters and *reaction* calorimeters.

Substance calorimeters comprise:

1. High-temperature heat capacity calorimeters (from room temperature upward) are used to obtain C_p°, from which both $H^\circ(T) - H^\circ(298.15)$ and $S^\circ(T) - S^\circ(298.15)$ can be obtained. Differential scanning calorimetry (DSC) is a relatively modern, easy-to-use technique which can be used for measuring C_p° on a few milligrams of material, although not with the accuracy achievable from using a calorimeter designed specifically for the purpose.

In addition, $H°(T) - H°(298.15)$ can be measured directly in a drop calorimeter, where the specimen at temperature T is dropped into a calorimeter maintained at room temperature. Many high-temperature $C_p°$ values have been obtained from the temperature derivative of the results obtained from using a drop calorimeter.

2. Low-temperature heat capacity calorimeters (from close to 0 K to room temperature) are used to obtain $C_p°$ and hence $S°(298.15 \text{ K})$ for all substances.

Reaction calorimeters, which are used to measure $\Delta_r H°$ at either 298.15 K or at elevated temperatures, can be further subdivided into *synthesis* calorimeters, in which the reactants combine to form the pure substance of interest, and *solution* calorimeters, in which $\Delta_f H°$ is obtained from dissolving both reactants and products in a suitable solvent and then using $\oint H° = 0$ to obtain the formation enthalpies from the solution enthalpies.

Although calorimetry is an important tool in obtaining the thermodynamic properties of pure condensed-phase substances, we will see later that it is also possible to obtain thermodynamic data for pure substances from phase and chemical equilibrium measurements.

We have also seen that it is important to have information for solids in their different structural forms. Many of these will be metastable or even unstable in certain temperature ranges and not amenable to direct measurement. In this case, the desired thermodynamic properties may have to be obtained by indirect methods.

In the last decade or so, the first-principles calculations of 0 Kelvin total energies have become commonplace. The best of the results from these density functional calculations are capable of giving results of comparable accuracy to those obtainable at 298.15 K by experimental measurement. Calculations have the big advantage over experimental measurement that they can be carried out on unstable structures just as readily as they can on the most stable. This is rarely possible experimentally.

The thermodynamic properties of perfect gases are usually not measured by any of the techniques outlined above but are obtained from a combination of spectroscopic measurements and statistical mechanics. This will be discussed briefly in Chapter 3.

EXERCISES

2.1   Show that the storage of data as the six coefficients mentioned in Table 2.1 for PbO leads to the following expression for the Gibbs energy:

$$G = A + BT + CT \ln(T) + DT^2 + ET^3 + \frac{F}{T} \qquad (2.46)$$

Relate the parameters in this equation to the heat capacity equation parameters given in (2.3).

2.2 (a) Calculate the enthalpy of oxidation of Si(l) to $SiO_2(\beta)$ at 1700 K using the data given below using both the standard substance and H^{SER} approaches. The result obtained from both methods should be identical.

(b) Calculate $\Delta_f G^\circ(SiO_2, 1700 \text{ K})$ using either of the reference states in (a) and compare your calculated result with that obtained using $\Delta_f H^\circ(SiO_2, 298.15\text{K}) = -910.7$ kJ mol^{-1} and the values of the gef at 1700 K.

(c) By assuming that $\Delta_f(\text{gef})$ is temperature independent, obtain the linear equation for $\Delta_f G^\circ$ as a function of temperature in the region of 1700 K.

The element Si melts at 1685 K and α-SiO$_2$ (alpha quartz) transforms to β-SiO$_2$ (beta quartz) at 844 K. Also β-SiO$_2$ is actually metastable above 1400 K, but this fact is to be ignored in the problem.

T/K	Substance	$(H_T^\circ - H_{298.15}^\circ)/T$ /J K^{-1} mol^{-1}	S_T° /J K^{-1} mol^{-1}	$-$gef /J K^{-1} mol^{-1}	$\langle C_p^\circ \rangle$ /J K^{-1} mol^{-1}
298.15	Si(s)	0	18.81	18.81	19.94
298.15	O$_2$(g)	0	205.15	205.15	29.37
298.15	SiO$_2(\alpha)$	0	41.46	41.46	44.59
1700	Si(l)	51.085	92.03	40.95	25.52
1700	O$_2$(g)	28.226	262.69	234.46	37.11
1700	SiO$_2(\beta)$	56.291	153.85	97.56	75.98

2.3 The element Fe, exhibiting body-centered-cubic (bcc) structure (α-Fe) at low temperatures, transforms first to face-centered-cubic (fcc) structure (γ-Fe) at 1042 K, next to the bcc structure (δ-Fe) at 1184 K, and to the liquid state (or melts) at 1665 K, with the pressure maintained at 1 bar.

The enthalpies of the transitions of Fe at these temperatures are 958.6, 899.8, and 1015.7 J mol^{-1} respectively.

The enthalpy of α-Fe at 1042 K relative to its enthalpy at 298.15 K is 26,410 J mol^{-1}.

The corresponding entropy changes are 0.92, 0.76, and 0.61 J K^{-1} mol^{-1}, respectively. The entropy of α-Fe at 1042 K is 68.84 J K^{-1} mol^{-1}.

The average specific heats $(\overline{C}_p)$ for γ-Fe (1042–1184 K), δ-Fe (1184–1665 K), and the liquid Fe, L-Fe (>1665 K) are 43.87, 35.71, and 41.69 J K^{-1} mol^{-1}, respectively.

(a) Calculate and plot the enthalpies of Fe as a function of temperature from 298.15 to 1800 K.

(b) Calculate and plot the entropies of Fe as a function of temperature from 298.15 to 1800 K.

(c) Calculate and plot the Gibbs energies of Fe as a function of temperature from 298.15 to 1800 K.

(d) We nearly always use 298.15 K as the reference state. However, it may become more convenient to use γ-Fe (1042 K) as the reference state. Calculate the Gibbs energies of Fe using this reference state and plot the Gibbs energies as a function of temperature from 298.15 to 1800 K. Convince yourself that the Gibbs energy difference between any chosen temperatures is the same regardless of the references state you use, that is, α-Fe at 298.15 K or γ-Fe at 1042 K.

2.4 Use some, if not all, of the following thermodynamic data to carry out the required calculations. The tabulated values refer to 1 mol of substance.

Substance	$\langle C \rangle_p^\circ/(\text{J K}^{-1})$	$\Delta_{\text{fus}} H^\circ/\text{J}$	$\Delta_f H^\circ(298 \text{ K})/\text{J}$	$S^\circ(1000 \text{ K})/(\text{J K}^{-1})$
Zn(s)	27.5	7,320	0	74.55
Zn(l)	31.38	—	—	86.54
$O_2(g)$	32.10	—	0	243.58
ZnO(s)	46.77	—	−350,460	101.76

Zinc melts at 693 K.

The oxidation of Zn(l) to ZnO(s) at 1000 K may be represented by the following chemical reaction: $\text{Zn(l)} + 0.5\,O_2(g) = \text{ZnO(s)}$.

(a) Based on the standard reference state, what are the enthalpies of Zn(l), $O_2(g)$, and ZnO(s) at 1000 K?

(b) Based on the SER state, what are the enthalpies of Zn(l), $O_2(g)$, and ZnO(s) at 1000 K?

(c) Using the SER state, calculate the enthalpy of oxidation of Zn(l) to ZnO(s) at 1000 K.

(d) Calculate the entropy of oxidation of Zn(l) at 1000 K.

(e) Calculate the Gibbs energy of oxidation of Zn(l) at 1000 K.

3 Calculation of Thermodynamic Properties of Unary Systems

The aim of microscopic thermodynamics or statistical mechanics is to provide an understanding, at the atomic level, of the observed macroscopic thermodynamic properties.

The calculation of ground-state energies and the energy levels of the various excited states for a substance is the business of quantum mechanics. The calculation of the occupation of these energy levels, and hence the mean energy and the spread in the energy, resulting in the calculation of the bulk properties of the substance, is the business of statistical mechanics.

In this Chapter we will attempt to illustrate the kind of things which result from the application of statistical mechanics without becoming too involved in the details. This will give rise to a few "it can be shown that..." statements.

Maxwell gave a thorough treatment of the kinetic theory of gases and showed how the bulk thermodynamic (and dynamic) properties of a gas could be calculated from its atomic properties. Boltzmann extended Maxwell's work and showed how probabilistic system averages could be obtained from kinetic (time averages)—an equivalence which is known as the ergodic hypothesis. He postulated that each microstate for a system has an equal probability of occurring and that the measured bulk properties are those obtained from the calculation of microstate probabilities. In particular, he related the bulk property S *for a system of fixed energy* to the number of accessible microstates (often called the thermodynamic probability W). Note that it is the number of accessible states and not the number of states actually accessed during the making of an observation. The number of accessed states is, generally, only a very small fraction of the number of possible accessible states.

The form of the Boltzmann relation can be understood by considering a system comprised of two subsystems. The entropy property is additive but the number of microstates is multiplicative:

$$S = S_1 + S_2 \qquad W = W_1 W_2 \tag{3.1}$$

From these differences in behavior, it follows that

$$S \ \alpha \ \log_e W \tag{3.2}$$

Materials Thermodynamics. By Y. Austin Chang and W. Alan Oates
Copyright 2010 John Wiley & Sons, Inc.

The proportionality constant is now known as the Boltzmann constant k_B and the final Boltzmann equation is then

$$\boxed{S = k_B \log_e W}\qquad(3.3)$$

This expression is often used, but it should be remembered that it applies to systems of fixed energy.

3.1   CONSTANT-PRESSURE/CONSTANT-VOLUME CONVERSIONS

Experimental measurements are usually carried out at constant pressure whereas theoretical calculations often refer to constant volume. It follows that we often need to be able to convert constant-volume properties to those at constant pressure. We illustrate this property transformation by considering the relation between the two heat capacities C_V and C_p. Macroscopic thermodynamics provides the necessary relationships for carrying out this and similar transformations.

If we select T and V as independent variables, then, from the differential form of the first law for $p-V$ work only,

$$\delta q = dU + p\,dV$$

$$\frac{\delta q}{\delta T} = \left(\frac{\partial U}{\partial T}\right)_V + \left[\left(\frac{\partial U}{\partial V}\right)_T + p\right]dV$$

$$= C_V + \left[\left(\frac{\partial U}{\partial V}\right)_T + p\right]dV\qquad(3.4)$$

If T and V are varied so that p remains constant, then, when T changes by dT, V will change by $(\partial V/\partial T)_p\,dT$ and the amount of heat transferred is

$$\delta q|_p = C_p\,dT = C_V\,dT + \left[\left(\frac{\partial U}{\partial V}\right)_T + p\right]\left(\frac{\partial V}{\partial T}\right)_p dT\qquad(3.5)$$

We now have the desired relation between the constant-pressure and constant-volume properties:

$$\boxed{C_p = C_V + \left(\frac{\partial V}{\partial T}\right)_p\left[\left(\frac{\partial U}{\partial V}\right)_T + p\right]}\qquad(3.6)$$

The experimental values for C_p° for Ti metal over a wide temperature range are shown in Figure 3.1. Not shown in the figure is the fact that $C_p^\circ \to 0$ as $T \to 0$ (the way it does this is discussed in Section 3.4). Titanium has been selected for illustration since it has an allotropic transformation and the figure shows how C_p° changes for the two solid phases, the liquid phase, and the gas phase [monatomic

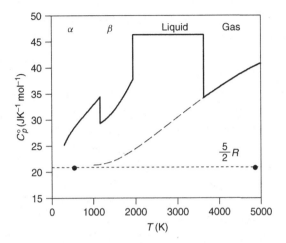

Figure 3.1 Standard heat capacity of Ti over wide temperature range.

Ti(g)]. From the Maxwell relation

$$\left(\frac{\partial C_p}{\partial p}\right)_T = -T\left(\frac{\partial^2 V}{\partial T^2}\right)_p \tag{3.7}$$

we see that, for perfect gases, C_p is pressure independent. For condensed phases, in the absence of any phase changes, (3.7) can be used to show that the effect of pressure on C_p is very small. It can also be seen in Figure 3.1 that, at each phase change, there is a discontinuity in C_p° resulting from the different ways in which the different energy levels are occupied in the various phases. The meaning of the dashed and dotted curves is discussed in Section 3.3.

The standard entropies of the various Ti phases are shown by the solid curves in Figure 3.2. Also shown are the nonstandard entropies at 0.01 bar. The properties of the condensed phases are unchanged by this pressure variation, but it can be seen that, contrary to the situation for C_p, S for the gas phase is pressure dependent, even when the gas is perfect. The entropy of a perfect gas increases as the pressure decreases. The transformation temperature between liquid and gas phases is also pressure dependent (the temperatures where the vapor pressure is 1 bar or 0.01 bar can be read from the graph). The large value for $\Delta_{vap}S^\circ$ as compared with $\Delta_\alpha^\beta S^\circ$ and $\Delta_{fus}S^\circ$ is also clearly apparent in this figure.

3.2 EXCITATIONS IN GASES

3.2.1 Perfect Monatomic Gas

(a) *Translational Contributions*   The following is based on the use of classical mechanics, not quantum mechanics, this being satisfactory for gases except at the lowest temperatures ($\lesssim 10$ K).

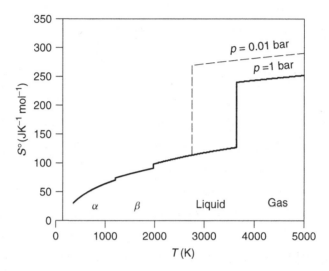

Figure 3.2   Solid curve: standard entropy of Ti; dashed curve: entropy at 0.01 bar.

For a system of N noninteracting particles of mass m confined in a box of volume V and traveling with root-mean-square velocity u, it can be shown that the pressure p is given by

$$p = \frac{1}{3}\frac{Nmu^2}{V} \qquad (3.8)$$

and since the kinetic energy (KE) of the particles $= 1/2 \; Nmu^2$, we see that

$$pV = \tfrac{2}{3}\text{KE} \qquad (3.9)$$

Since there is no other contribution to the energy of the noninteracting particles, KE is the same as the total energy U, and since the EOS is that for a perfect gas, $pV = Nk_BT$, we obtain

$$\boxed{U_m = \tfrac{3}{2}RT} \quad \text{(noninteracting particles)} \qquad (3.10)$$

The three degrees of freedom for a three-dimensional translator mean that there is an energy contribution of $1/2RT$ per degree of freedom. This theorem of the *equipartition of energy* (Boltzmann) applies to all other possible degrees of freedom where classical mechanics is valid.

From (3.10) we obtain the following equation for the constant-volume heat capacity:

$$\boxed{C_V = \left(\frac{\partial U_m}{\partial T}\right)_V = \frac{3}{2}R} \quad \text{(noninteracting particles)} \qquad (3.11)$$

We can now use (3.3) to obtain C_p. For a perfect gas of noninteracting particles $(\partial U / \partial V)_T = 0$ and, using the perfect gas law, we obtain

$$C_p = C_V + \left(\frac{R}{V}\right) V = C_V + R = \tfrac{5}{2} R \tag{3.12}$$

Note that this value is pressure independent, in agreement with the results for Ti(g) shown in Figure 3.1. It can also be seen in this figure that this value is found for Ti(g) at low temperatures but not at high temperatures (the explanation for this is given below).

We cannot obtain S from C_p just by using $S = \int C_p / T \, dT$ since the allowed energies for particles in a box depend not just on temperature but also on the volume of the box. Without going into the derivation, the *Sackur–Tetrode* equation gives the translational entropy for a gas of noninteracting particles as a function of p and T:

$$\frac{S_m}{R} = \frac{5}{2} \ln T - \ln p + \ln \left(\frac{2\pi m}{h^2}\right)^{3/2} k_B^{5/2} + \frac{5}{2} \tag{3.13}$$

The first term in this expression leads to $C_p = 5/2R$ and it can be seen that, as shown in Figure 3.2, the translational entropy increases as the pressure decreases.

(b) *Electronic Contributions*   The reason that, at high temperatures for monatomic gases, $C_p \neq 5/2R$ (see Fig. 3.1) is where electronic excitations are being superimposed on the translational component of the monatomic gas: There are many electronic energy levels and these become increasingly occupied with increasing temperature, that is, the energy spread increases.

3.2.2 Molecular Gases

Molecular gases have extra modes for absorbing energy so that, as well as translation and electronic, other contributions to C_p have to be considered, namely, rotation and harmonic and anharmonic vibration. The energy levels associated with these excitations can be obtained from spectroscopic measurements and the levels used in the appropriate equations derived from statistical mechanics for calculating C_p° and S°. This combination of spectroscopic-level measurements with theory is now the accepted method for obtaining the best values of standard thermodynamic functions for perfect gases.

3.3 EXCITATIONS IN PURE SOLIDS

The thermal excitations in the energy for elemental or pure compound solids are not translational, as in gases, but involve vibrational, electronic, and magnetic contributions:

(a) *Vibrational*   Application of the Boltzmann energy equipartition theorem to a solid, which possesses six vibrational degrees of freedom, leads to *Dulong and Petit*'s equation for the constant-volume heat capacity, namely, $C_V = 6 \times 1/2R = 3R = 24.95$ J K^{-1} mol^{-1}. It is apparent, however, from experimental results, that this equation does not hold at low temperatures, as can be seen in Figure 3.3, which shows the standard constant-pressure heat capacity of silver (the constant-pressure/volume correction is quite small in this region so that $C_V \sim C_p$).

An explanation for the principal features of the low-temperature heat capacity of solids was one of the early triumphs in the application of quantum theory.

Einstein considered a system of N atoms as a set of $3N$ independent simple harmonic oscillators with angular frequency ω. From quantum mechanics, the energy levels of such an oscillator are given by

$$E_n = (n + \tfrac{1}{2})\hbar\omega \tag{3.14}$$

where n is the principal quantum number and the mean or expectation value (the bulk observed value) of the energy is related to the probabilities of the occupation of the different energy levels:

$$\langle E \rangle = \sum_n p_n E_n \tag{3.15}$$

The calculation of the probabilities is carried out using statistical mechanics and this mean energy is found to be

$$\langle E \rangle = \frac{1}{2}\hbar\omega + \frac{\hbar\omega}{\exp[\hbar\omega/(kT)] - 1} \tag{3.16}$$

Since the oscillators are assumed independent, the molar internal energy is given by $3N$ times this and the constant-volume heat capacity is then found, by differentiation, to be

$$C_V = 3R\left(\frac{\theta_E}{T}\right)^2 \frac{\exp(\theta_E/T)}{\left[\exp(\theta_E/T) - 1\right]^2} \tag{3.17}$$

where the Einstein temperature $\theta_E = \hbar\omega/k_B$.

This equation produces a curve of similar form to that shown for Ag in Figure 3.3. It also gives the desired classical high-temperature limit of $3R$. The weakness of the Einstein model lies in the assumption of independent oscillators vibrating with a single frequency. The atoms in real solids are strongly coupled with a continuous spectrum of vibration frequencies. *Debye* removed the Einstein assumption and made the simplest possible assumption for the vibrational frequency spectrum (a continuous distribution with maximum frequency ω), from which he was able to derive the following equation for C_V in terms of a single parameter, the Debye temperature $\theta_D = \hbar\omega/k_B$:

$$C_V = 3R\left[4D\left(\frac{\theta_D}{T}\right) - \frac{(3\theta_D/T)}{\exp(\theta_D/T) - 1}\right] \tag{3.18}$$

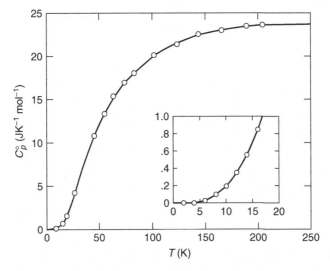

Figure 3.3   Constant-pressure heat capacity of Ag(s) below room temperature.

where D is known as the Debye function defined by

$$D(x) = \frac{3}{x^3} \int_0^x \frac{z^3 \, dz}{\exp(z) - 1} \tag{3.19}$$

The Debye equation for C_V indicates that, if we plot C_V versus θ_D/T, the results for all solid substances lie on the same curve. This curve is shown in Figure 3.4.

The vibrational contributions to the internal energy, entropy, and Helmholtz energy according to the Debye model are given by

$$U_m \text{ (Debye)} = \tfrac{9}{8} R\theta_D + 3RTD\left(\frac{\theta_D}{T}\right) \tag{3.20}$$

$$S_m \text{ (Debye)} = R\left[-3 \log_e(1 - \exp(-\frac{\theta_D}{T} + 4D\left(\frac{\theta_D}{T}\right)\right] \tag{3.21}$$

$$A_m \text{ (Debye)} = \tfrac{9}{8} R\theta_D + RT\left[3 \log_e(1 - \exp(-\frac{\theta_D}{T} - D\left(\frac{\theta_D}{T}\right)\right] \tag{3.22}$$

Here, the first, temperature-independent, term in U_m and A_m represents the zero-point energy (ZPE), which originates from the atoms being in their lowest vibrational state at 0 K.

The Debye model turns out to be fairly satisfactory for metals and alloys, and we will use it again in Section 3.4 when discussing the temperature and volume variation of the thermodynamic properties of a solid.

(b) *Electronic*   Drude (1900) realized that the high electrical conductivity of metals was due to free electrons. But if classical Maxwell–Boltzmann statistics

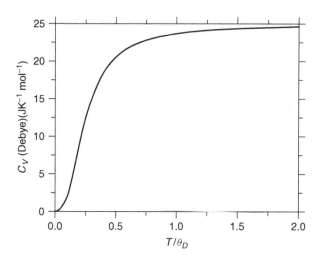

Figure 3.4 C_V for all metals fall approximately on this curve when the temperature scale is normalized in the manner shown.

are applied to this free gas, then, just as for the translation heat capacity of a monatomic perfect gas, it is expected that $C_V^{el} = 3/2R$. Such a high value is inconsistent with the observed high-temperature heat capacity of metals, which are approximately $3R$. It is known, as discussed above, that this value arises from lattice vibrations, whereas a value of $9/2R$ would result if we were to add the classic translation value to the vibrational contribution. It was again left to quantum mechanics to provide the answer. Sommerfeld applied the Pauli exclusion principle (one electron only in each electron energy level), which leads to all the lower electron levels being fully occupied. As a result, only those electrons near the Fermi surface, which separates the occupied from the unoccupied levels, can be excited to higher levels. It is this which is responsible for the low values of the electron heat capacity of solids. It can be shown that

$$C_V^{el} = \tfrac{1}{3}(\pi k_B)^2 N(E_F)T = \gamma_{el}T \qquad (3.23)$$

where $N(E_F)$ is the density of states at the Fermi level. The electronic specific heat coefficient γ_{el} lies in the range of approximately $1\text{--}30.10^{-4}$ J K^{-2} mol^{-1}. This leads to a value of $C_V^{el} << C_V^{vib}$ except at temperatures close to 0 K.

(c) *Magnetic* Ferromagnetism is associated with the long-range ordering of the atomic electron spins in a metal. Paramagnetism results when these spins are almost completely randomly distributed. Several pure substances undergo a transformation from the ferromagnetic to the paramagnetic state. The most notable examples are the metals Fe, Co, and Ni. The transformation from ordering to disordering of the spins means that the populated energy levels change with temperature and this results in a heat capacity change. The transformation from ordered to disordered spins is a second-order one (discussed in Chapter 15), for

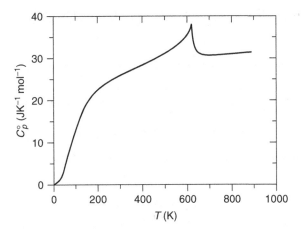

Figure 3.5 Heat capacity for Ni, showing a cusp associated with magnetic contribution.

which the heat capacity–temperature curve shows a characteristic cusp like that shown for Ni in Figure 3.5. The temperature where there is a sudden drop in C_p is known as the *Curie* temperature.

Note that some substances are anti-ferromagnetic, where the low-temperature ordered state consists of the spins being antiparallel, so that there is no net magnetic moment in the fully ordered state (the spin alignment can be observed in neutron-scattering experiments). The transition to the disordered paramagnetic state occurs at the *Néel* temperature.

3.4 THE THERMODYNAMIC PROPERTIES OF A PURE SOLID

In the above discussion we have concentrated principally on the heat capacities of a pure solid. We now wish to concentrate on the other thermodynamic properties of a simple pure solid. In order to do so, we will consider the simplest possible model in some detail in order to illustrate the approach. The model has been chosen to simplify the mathematics without making the physics unrealistic:

(i) Only vibrational excitations are considered and we assume that these can be represented by the Debye model. The Debye temperature is assumed to be independent of both temperature and volume.

(ii) The thermal expansivity and compressibility are assumed to be independent of temperature and volume.

The volume coefficient of expansion, α, and the isothermal compressibility, κ_T, are defined as

$$\alpha = \frac{1}{V_m}\left(\frac{\partial V_m}{\partial T}\right)_p \qquad \kappa_T = -\frac{1}{V_m}\left(\frac{\partial V_m}{\partial p}\right)_T \qquad (3.24)$$

If we consider a differential volume change as a function of pressure and temperature,

$$dV = \left(\frac{\partial V}{\partial T}\right)_p dT + \left(\frac{\partial V}{\partial p}\right)_T dp$$

By introducing our model approximations into this equation, we obtain

$$dV = V_0^{\,0}(\alpha\, dT - \kappa_T\, dp) \tag{3.25}$$

which, on integration, gives the following simple EOS for our solid:

$$V = V_0^{\,0}(1 + \alpha T + \kappa_T p) \tag{3.26}$$

$$= V_T^{\,0}\kappa_T p \tag{3.27}$$

where $V_0^{\,0}$ is the volume at zero temperature and zero pressure and $V_T^{\,0}$ is the volume at temperature T and zero pressure.

We may use Maxwell relations in order to obtain the volume derivatives of the thermodynamic properties from the EOS. The following table shows the general Maxwell relations, applicable to any EOS, and also the specific values for the EOS being used:

Derivative	General	Specific
$\left(\dfrac{\partial U}{\partial V}\right)_T$	$T\left(\dfrac{\partial p}{\partial T}\right)_V - p$	$\dfrac{\alpha T}{\kappa_T} - p$
$\left(\dfrac{\partial S}{\partial V}\right)_T$	$\left(\dfrac{\partial p}{\partial T}\right)_V$	$\dfrac{\alpha}{\kappa_T}$
$\left(\dfrac{\partial A}{\partial V}\right)_T$	$-p$	$V_T^{\,0}$

We may now use these relations to obtain the *static* (abbreviation stat) contributions to the thermodynamic properties:

$$U_{\text{stat}}(V, T) - U^{\circ}(V_T^{\,0}) = \frac{\alpha T}{\kappa_T}\int_{V^0}^{V} dV - \int_{V^0}^{V} p\, dV \tag{3.28}$$

$$= \frac{\alpha T}{\kappa_T}(V - V_T^{\,0}) + \int_{V^0}^{V}\frac{V - V_T^{\,0}}{\kappa_T V_0^{\,0}}\, dV \tag{3.29}$$

$$= \frac{\alpha T}{\kappa_T}(V - V_T^{\,0}) + \frac{\frac{1}{2}(V - V_T^{\,0})^2}{\kappa_T V_0^{\,0}} \tag{3.30}$$

$$S_{\text{stat}}(V, T) - S^{\circ}(V_T^{\,0}) = \frac{\alpha}{\kappa_T}\int_{V_T^0}^{V} dV \tag{3.31}$$

$$= \frac{\alpha}{\kappa_T}(V - V_T^0) \tag{3.32}$$

$$A_{\text{stat}}(V, T) - A^{\circ}(V_T^0) = \frac{\frac{1}{2}(V - V_T^0)^2}{\kappa_T V_0^0} \tag{3.33}$$

From the Maxwell relation,

$$\left(\frac{\partial A}{\partial V}\right)_T = -p \tag{3.34}$$

we see that the minimum in A with respect to V occurs at $p = 0$.

Using the Debye model for the vibrational contributions, the volume-independent variation of $U^{\circ}(V_T^0)$ and $A^{\circ}(V_T^0)$ are given by (3.20) and (3.22). We now have the necessary information to be able to calculate the variation of these properties as a function of both V and T. This variation is shown in Figures 3.6 and 3.7 for a particular set of values for θ_D, α, and κ_T. The values used ($\theta_D = 250$ K, $\alpha = 10^{-4}$ K^{-1}, $\kappa_T = 10^{-11}$ Pa^{-1}, $V_0^0 = 10^{-5}$ m^3 mol^{-1}, are not unreasonable values for these properties of real solids, although they have been chosen so as to emphasize the following effects shown in the figures:

(i) The vertical positioning of the curves is determined by θ_D.
(ii) The minima in the plots of A_m versus V/V_0^0 occur at $p = 0$. The variation of the position of these minima with temperature is determined by α.
(iii) The variation in U_m and A_m with V/V_0^0 at constant temperature is determined by κ_T.

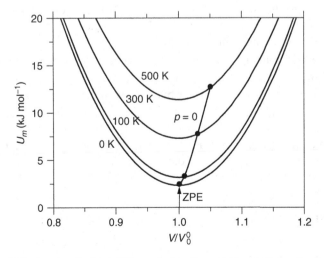

Figure 3.6 Calculated variation of U_m as a function of V and T for the simple model discussed in the text.

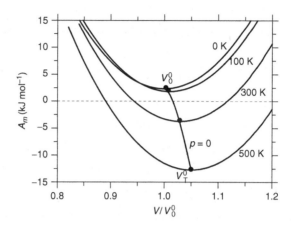

Figure 3.7 Calculated variation of A_m as a function of V and T for the simple model discussed in the text.

Similar calculations of H and G, using T and p as independent variables, are possible.

3.4.1 Inadequacies of the Model

As we mentioned earlier, the EOS used in the above calculations is too simple to represent the $p-V-T$ properties of a real solid. In particular, we should note that:

(a) The Debye model has its own limitations. For example, different θ_D are obtained from the equations for C_V(Debye), S(Debye), and A(Debye) and the temperatures obtained are found to vary with T, even when anharmonic contributions are insignificant. The Debye model does not, of course, allow for any such anharmonic contributions.

(b) If the Debye model is to be used, it is necessary to allow for the variation of θ_D.

(c) The α values are not constant but vary with T, going to zero at 0 K.

(d) The κ_T values are not constant but vary with pressure and temperature.

These and other factors complicate the calculation of the thermodynamic properties of pure solids. Nevertheless, our simple model gives a good introduction to the kind of approach which is necessary.

EXERCISES

3.1 Use Figure 3.3 to estimate the Debye temperature for Ag as a function of temperature. You will find a set of Debye Tables, available on the internet, most useful for this purpose. Compare the values of the Debye temperature obtained with values found in the literature.

3.2 At low temperatures, the Debye equation gives a T^3 relation $C_V = aT^3$, where $a = (12/5)(\pi^4 R/\theta_D^3)$. Given the following heat capacity data for Ag at low temperatures, obtain a value for θ_D and for the electronic heat capacity coefficient, γ.

T/K	C_p/J K-1 mol^{-1}	T/K	C_p/J K-1 mol^{-1}	T/K	C_p/J K-1 mol^{-1}
1	7.0×10^{-4}	15	0.682	75	17.2
2	2.6×10^{-4}	20	1.71	100	20.2
3	6.6×10^{-4}	25	3.12	150	23.1
4	1.36×10^{-3}	30	4.73	200	24.3
5	2.49×10^{-3}	40	8.24	250	25.0
10	1.88×10^{-2}	50	11.6	298	25.4

3.3 Show that the relation

$$C_p - C_V = \left(\frac{\partial V}{\partial T}\right)_T \left[\left(\frac{\partial U}{\partial V}\right)_p + p\right] \tag{3.35}$$

can be reduced to the form

$$C_p - C_V = \frac{\alpha^2 V_m T}{\kappa_T} \tag{3.36}$$

(Hint: take advantage of the Maxwell relations given in Section 1.1.5.1)

3.4 Show that $C_p - C_V = R$ for a perfect gas.

3.5 Show thermodynamically that $\kappa_T/\kappa_S, = C_V/C_p$.

3.6 When U is written as a function of pV/R for an ideal gas, show that $(\partial U/\partial p)_V = VC_V/R$.

3.7 At room temperature, taken as 298 K, Al has the following properties: $A_r = 26.9815$, density $= 2.70$ g cm^{-3}, $C_p = 24.31$ J mol^{-1}, $\alpha = 71.4 \times 10^{-6}$ K^{-1}, $\kappa_T = 1.34 \times 10^{-2}$ GPa^{-1}, $\theta_D = 375$ K.

Calculate (a) C_V and (b) κ_S. (c) An equation which has been used for the quick estimation of high-temperature heat capacities from ones at room temperature is the Nernst–Lindemann equation $C_p - C_V$ (Dulong–Petit) $= AC_p^2 T$, where the constant A in the equation is obtained from room temperature values.

Calculate C_p for Al(s) from $\theta_D/2$ to 933.25 K (the melting point). Compare the calculated results with any experimental data you find in the literature and offer comments on the value of the Nernst–Lindemann equation.

4 Phase Equilibria in Unary Systems

The calculation of phase diagrams provides an excellent illustration of the way in which thermodynamics permits one set of properties to be calculated from an apparently completely different set of properties.

For the moment we are concentrating on unary (one-component, pure substance) systems. As we will see, however, many of the features which are encountered in binary phase diagrams have their counterpart in unary phase diagrams.

In the absence of external fields (e.g., magnetic), the *state* of a one-component system is uniquely specified by two variables (see Section 2.3). Different choices of these variables are possible and give different possibilities for a two-dimensional representation of the system's phase behavior (phase diagrams). We have mentioned previously that the intensive thermodynamic variables can be divided into two categories: *field* variables and *density* variables. A field variable, for example, T, takes on identical values in any coexisting phases at equilibrium, while a density variable, for example, V_m, does not.

Figure 4.1 shows examples of the only possible different types of two–dimensional phase diagrams for a unary system:

(A) *Field–Field Type* The $p-T$ (field–field) phase diagram (Fig. 4.1a) shows single-phase areas, two-phase coexistence lines, a triple point, and a *critical point*. A two-phase line indicates where two coexisting phases have the same temperature and pressure. Similarly, at the *triple point*, the coexisting gas, liquid, and solid phases have exactly the same temperature and pressure. At the critical point the liquid and gas phases become indistinguishable.

(B) *Field–Density Type* In the $p-S_m$ (field–density) representation (Fig. 4.1b) there are single-phase areas as in a field–field diagram. But two-phase coexistence is now indicated by areas, and the triple point has become a horizontal triple line. The horizontal tie-lines in, say, the gas–liquid coexistence region indicate that any coexisting gas and liquid must have the same pressure (the field variable), but they have different S_m (the density variable). A similar interpretation holds for the

Materials Thermodynamics. By Y. Austin Chang and W. Alan Oates
Copyright 2010 John Wiley & Sons, Inc.

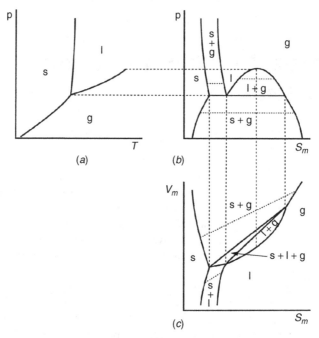

Figure 4.1   Different types of phase diagram for one-component system: (*a*) field–field; (*b*) field–density; (*c*) density–density.

gas, liquid, and solid coexisting along the triple line—equality of p but different S_m.

(C) *Density–Density Type*   In the V_m–S_m representation shown in Figure 4.1c, we have two-phase regions with oblique tie-lines and a triple triangle. The sloping tie-lines in the gas–liquid region, for example, tell us that any coexisting gas and liquid have different V_m as well as different S_m (although they still have the same p and T). The usual lever rule applies to the determination of the relative amounts of any coexisting phases along one of these sloping tie-lines. Any point in the triple triangle represents a system with coexisting gas, liquid, and solid phases, which differ from each other in both V_m and S_m. The lever rule can also be applied in this three-phase region for the determination of the relative volume amounts of the three phases.

When the solid phase has more than one crystal structure (polymorphism or allotropy), then unary phase diagrams can become more complex. A p–T phase diagram for pure Fe is shown in Figure 4.2. At $p \rightarrow 0$ we see the usual transformations bcc $\rightarrow$ fcc $\rightarrow$ bcc $\rightarrow$ liquid as the temperature is increased, but at high p it can be seen that the pattern of transformations is quite different.

A plot of one thermodynamic function against another does not always give rise to a *true* phase diagram, defined as one where any point on the diagram gives

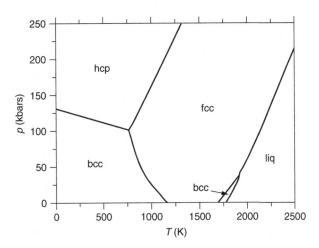

Figure 4.2 $p-T$ phase diagram for Fe.

a *unique* answer concerning the existence of individual phases and combinations of phases. The ones shown in Figure 4.1 are true phase diagrams, but some diagrams which show the range of existence of phases and their combinations are not. This is illustrated in Figures 4.3a and 4.3b. Although these diagrams are field–density diagrams and the topography of Figure 4.3a is similar to that for the plot of p versus S_m shown in Figure 4.3b, it is *not* a true phase diagram. This is apparent from Figure 4.3b, which shows a similar p-versus-V_m diagram but where the molar volume of the solid is greater than that of the liquid. Water and bismuth are common examples of materials which fall into this latter category. It is clear that in the neighborhood of the black dot the system's properties are not uniquely defined.

Plotting *conjugate variables*, for example, p versus V_m, is the source of the problem. Any plot of a property versus its conjugate property does not give a true phase diagram. We will meet this again when discussing binary systems (see

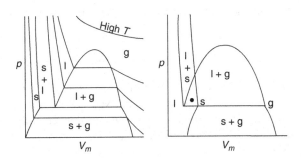

Figure 4.3   Plots of conjugate variables are not true phase diagrams: (*a*) appears to be a true phase diagram when $V_m(l) > V_m(s)$; (*b*) clearly not a true phase when $V_m(s) > V_m(l)$.

Chapter 8). Suffice to mention here that the definition of conjugate variables is somewhat broader than encountered previously. For example, it is possible to rewrite the fundamental equation as

$$dS_m = \frac{1}{T}dU_m + \frac{p}{T}dV_m \qquad (4.1)$$

In this case, p/T and V_m are seen to be conjugate variables so that a plot of these quantities will not give a true phase diagram.

An interesting feature arises for properties in the liquid–gas region. It is found that, if the $p-V-T$ variables are expressed relative to the values at the critical point for that particular substance, that is, as reduced variables

$$p_r = \frac{p}{p_c} \qquad V_r = \frac{V}{V_c} \qquad T_r = \frac{T}{T_c} \qquad \rho_r = \frac{\rho}{\rho_c} \qquad (4.2)$$

then the liquid–vapor coexistence curves plotted in terms of these reduced variables *approximately* coincide for simple molecules. These are examples of the application of the *law of corresponding states*, which can be useful in the estimation of the properties for substances where only limited information is available.

4.1 THE THERMODYNAMIC CONDITION FOR PHASE EQUILIBRIUM

Since we are usually interested in phase equilibrium at constant total pressure p_{total} and T, the following derivation is based on using the Gibbs energy, whose natural variables are p and T.

Consider two phases, liquid and gas, to be in open contact as illustrated in Figure 4.4. Matter can be transferred between the phases at constant p_{total} and T, although the two-phase system is considered closed.

The Gibbs energy of the whole system, comprising the two phases, is then

$$G^{total} = G^{liq} + G^{gas} \qquad (4.3)$$

and we wish to find the minimum in this total Gibbs energy, which we can write as

$$G^{total} = n_A^l \mu_A^l + n_A^g \mu_A^g \qquad (4.4)$$

There is, however, a constraint on this minimization, since the complete liquid/gas system is closed and the total amount of a is fixed, $n_A^{total} = n_A^l + n_A^g$.

Constrained minimization problems like this are conveniently handled using Lagrangian multipliers (see Appendix B for an introduction). In this case, only one λ is required:

$$\lambda : \qquad n_A(total) - n_A^l - n_A^g = 0 \qquad (4.5)$$

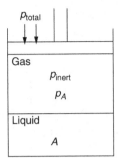

Figure 4.4 Phase equilibrium between pure liquid A and a gas phase comprising A and an inert gas. Both gas and liquid are maintained at the same total pressure p_{total}.

and the Lagrangian function $\mathcal{L}$ to be minimized becomes

$$\mathcal{L} = G + \lambda(n_A - n_A^l - n_A^g) \tag{4.6}$$

$$= n_A^l \mu_A^l + n_A^g \mu_A^g + \lambda(n_A - n_A^l - n_A^g) \tag{4.7}$$

Partial differentiation with respect to n_A^l and n_A^g gives the minimum in $\mathcal{L}$:

$$\frac{\partial \mathcal{L}}{\partial n_A^l} = \mu_A^l - \lambda = 0 \tag{4.8}$$

$$\frac{\partial \mathcal{L}}{\partial n_A^g} = \mu_A^g - \lambda = 0 \tag{4.9}$$

Elimination of the Lagrangian multiplier leads to the condition for the constrained minimum in G, that is, the condition of equilibrium:

$$\boxed{\mu_A^l = \mu_A^g} \tag{4.10}$$

As is apparent from (4.8) and (4.9), the physical meaning of the Lagrangian multiplier used for the mass balance constraint, λ, is seen to be that of the chemical potential.

For either of the single phases we can write

$$G(A) = n_A \mu_A \tag{4.11}$$

or

$$\mu_A = \frac{G(A)}{n_A} = G_m(A) \tag{4.12}$$

that is, for a unary system, we may also write the phase equilibrium condition as

$$G_m \left(A^l, p, T \right) = G_m \left(A^g, p, T \right) \tag{4.13}$$

Equations (4.10) and (4.13) appear to be equivalent ways of expressing the condition of phase equilibrium. There is, however, an important difference between these two equations. Equation (4.10) is still obtained if we start with different fixed external conditions; for example, at constant S and V and we used the fundamental equation for U, whereas (4.13) applies only at constant p and T.

The conditions for two bulk phases of a pure substance to be in equilibrium then are:

(i) Their temperatures must be equal so that there is no heat flow between the two phases.
(ii) Their total pressures must be equal so that there is no bulk material flow between the two phases.
(iii) The chemical potentials of the substance must be equal in the two phases so that there is no mass (diffusive) flow between the two phases.

As will become apparent later, this last criterion, as well as the usual equality of p and T, holds in much more complex situations than those being discussed here.

It is very important to appreciate that this equilibrium condition for mass transfer, (4.10), is a *necessary but not sufficient condition for evaluating the global equilibrium* in a system. The condition (4.10) alone can only be used when the two phases are actually specified beforehand, for example, A^l and A^g. These, however, may not be the global equilibrium ones for the specified conditions. For example, in Figure 4.4, the equilibrium condensed phase might be solid a rather than liquid a under the specified p and T. At this total p and T, we can use (4.10) to evaluate p_A in the gas phase for equilibrium with either liquid or solid a. Under the particular conditions chosen, only one of them would represent the global equilibrium. The other would represent a metastable equilibrium. It is necessary to check, using (4.4), which of the two equilibria has the lower Gibbs energy in order to decide which is the global equilibrium and which is a local, metastable, equilibrium.

4.2 PHASE CHANGES

4.2.1 The Slopes of Boundaries in Phase Diagrams

Consider a $p-T$ plot for a pure substance like that shown in Figure 4.1a. An important relation for the slopes of the two-phase equilibrium lines on such a diagram can be derived in the following way.

At any point along the line for solid–liquid equilibrium

$$G(s) = G^l \tag{4.14}$$

Since this applies for all points along the two-phase equilibrium curve, then for any displacement from a particular point along the curve for two-phase equilibrium

$$\Delta[G(l) - G(s)] = 0 \tag{4.15}$$

For one of the phases, the total differential for G as a function of p and T is

$$dG = \left(\frac{\partial G}{\partial p}\right)_T dp + \left(\frac{\partial G}{\partial T}\right)_p dT \tag{4.16}$$

Repeating for the other phase and substituting into (4.15) and rearranging give the slope of the pressure–temperature phase equilibrium curve:

$$\frac{dp}{dT} = -\frac{\Delta\left[(\partial G/\partial T)_p\right]}{\Delta\left[(\partial G/\partial p)_T\right]}$$

$$= \frac{\Delta S}{\Delta V} \tag{4.17}$$

and since $\Delta G = 0$ along the two-phase equilibrium, $\Delta H = T\,\Delta S$. Hence, we may rewrite (4.17) as

$$\boxed{\frac{dp}{dT} = \frac{\Delta H}{T\,\Delta V}} \tag{4.18}$$

This last equation is known as the *Clausius–Clapeyron* equation. From it, we can predict the slope of a p–T phase boundary from the enthalpy and volume change associated with the phase transformation. Alternatively, the inverse of (4.18) can be used to obtain the variation of $\Delta_{fus}H_m^*$ as a function of pressure from measurements of the pressure change of the melting point and volume. These represent good examples of using thermodynamic relations for predicting one property of a substance from properties which might be easier to measure.

It is found experimentally that, while $\Delta_{fus}V_m^*$ and T_{fus} change markedly with pressure, the change in $\Delta_{fus}H_m^*$ is small, that is,

$$\Delta_{fus}H_m^* \approx \Delta_{fus}H_m^\circ \tag{4.19}$$

A special case of the Clausius–Clapeyron equation, which involves using *approximations*, arises in its application to gas/condensed-phase equilibrium at low pressures. Assuming the gas is perfect (first approximation) and that

$V_m(\text{l}) \ll V_m^{\text{g}}$ (second approximation), then almost all the volume change occurs in the gas phase and is given by

$$\Delta_{\text{vap}} V \approx \frac{RT}{p} \tag{4.20}$$

giving

$$\boxed{\frac{d \log_e p}{dT} \approx \frac{\Delta_{\text{vap}} H}{RT^2}} \qquad \text{(liquid (or solid), perfect gas only)} \tag{4.21}$$

which is known as the *Clapeyron* equation.

Another aspect of the slopes of phase boundaries which is apparent in field–field phase diagrams is concerned with how two-phase equilibrium curves extend through three-phase equilibrium points (see Fig. 4.5a).

Consider the solid–liquid–gas equilibrium shown in the sketch. Using the Clausius–Clapeyron equation, with superscripts to indicate which two-phase equilibrium is being considered, we see that

$$\left(\frac{dp}{dT}\right)^{\text{s-g}} = \frac{\Delta_{\text{s}}^{\text{g}} H}{T \Delta_{\text{s}}^{\text{g}} V} = \frac{\Delta_{\text{l}}^{\text{g}} H + \Delta_{\text{s}}^{\text{l}} H}{T \Delta_{\text{s}}^{\text{g}} V} \tag{4.22}$$

$$= \frac{\Delta_{\text{l}}^{\text{g}} V}{\Delta_{\text{s}}^{\text{g}} V} \left(\frac{dp}{dT}\right)^{\text{l-g}} + \frac{\Delta_{\text{s}}^{\text{l}} V}{\Delta_{\text{s}}^{\text{g}} V} \left(\frac{dp}{dT}\right)^{\text{s-l}} \tag{4.23}$$

$$= (1 - v_f) \left(\frac{dp}{dT}\right)^{\text{l-g}} + v_f \left(\frac{dp}{dT}\right)^{\text{s-l}} \tag{4.24}$$

where v_f is the volume fraction of the total s–g volume change which is due to the s–l volume change.

It follows from (4.24) that the slope of the s–g curve must lie between that of the s–l and l–g curves; that is, the extension of all three curves through

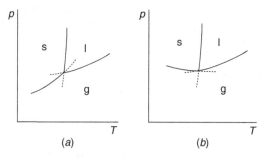

Figure 4.5 Extensions of phase boundaries in field–field phase diagrams: (a) correct; (b) impossible.

the triple point is into the opposite single-phase region. This phase boundary extension requirement is known as *Schreinemaker's rule*.

An impossible extension is shown in Figure 4.5b.

The equivalent requirement of Schreinemaker's rule in field–density and density–density phase diagrams such as those shown in Figures 4.1b and 4.1c is that the phase boundaries must always extend into the two-phase regions on the opposite side of the invariant point. Some published phase diagrams indicate that this thermodynamic requirement has not always been complied with by experimentalists.

4.2.2 Gibbs Energy Changes for Phase Transformations

The magnitudes of the Gibbs energies for any phase change in pure substances is of paramount importance. We will meet these quantities frequently in the application of thermodynamics to the calculation of binary phase diagrams (see Chapters 8 to 12).

At the equilibrium transition temperature for a pure substance at constant pressure

$$\Delta_{\mathrm{tr}} G_m^*(T_{\mathrm{tr}}^*) = 0 \tag{4.25}$$

and so

$$\Delta_{\mathrm{tr}} S_m^*(T_{\mathrm{tr}}^*) = \frac{\Delta_{\mathrm{tr}} H_m^*(T_{\mathrm{tr}}^*)}{T_{\mathrm{tr}}^*} \tag{4.26}$$

It can be seen from the sketches in Figure 4.6, which apply for standard conditions, that $\Delta_{\mathrm{tr}} G_m^* \neq 0$ at temperatures away from the equilibrium transition temperature.

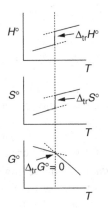

Figure 4.6 Variation of enthalpy, entropy, and Gibbs energy differences in neighborhood of first-order phase transformation.

A linear approximation for $\Delta_{tr}G^\circ$, which is equivalent to assuming that ΔH° and ΔS° are temperature independent, is also very useful and much used:

$$\Delta_{tr}G^\circ(T) = \Delta_{tr}H^\circ(T) - T\Delta_{tr}S^\circ(T) \tag{4.27}$$

$$0 = \Delta_{tr}H^\circ(T_{tr}^\circ) - T_{tr}^\circ \Delta_{tr}S^\circ(T_{tr}^\circ) \tag{4.28}$$

These two equations always hold, but if we now assume that $\Delta_{tr}H^\circ(T) \approx \Delta_{tr}H^\circ(T_{tr}^\circ)$ and $\Delta_{tr}S^\circ(T) \approx \Delta_{tr}S^\circ(T_{tr}^\circ)$, then a useful approximation is obtained:

$$\boxed{\Delta_{tr}G^\circ(T) \approx (T_{tr}^\circ) - T)\Delta_{tr}S^\circ \approx \frac{(T_{tr}^\circ) - T}{T_{tr}^\circ}\Delta_{tr}H^\circ(T_{tr}^\circ)} \tag{4.29}$$

Some useful approximations for the magnitudes of $\Delta_{tr}S^\circ(T_{tr})$ and $\Delta_{tr}H^\circ(T_{tr})$ for various types of phase transformation in pure substances are summarized below. Approximations like these are useful when it is necessary, due to the absence of experimental data, to estimate properties.

(A) *Evaporation*   Since $S^{\circ,g} \gg S^{\circ,l}$, then it follows that $\Delta_{vap}S^\circ \approx S^{\circ,g}$ and since all monatomic gases have approximately the same entropy (see Chapter 3), it comes as no surprise that $\Delta_{vap}S_m^\circ$ is approximately constant for all metals (very few metals have significant amounts of molecular species in their vapors). This is the basis of *Trouton's rule*:

$$\Delta_{vap}S_m^\circ \approx 84 \text{ J mol}^{-1} \text{ K}^{-1} \tag{4.30}$$

If the transformation temperature is 1000 K, we see that $\Delta_{vap}H_m^\circ$ is estimated to be ≈ 84 kJ mol^{-1}, which is quite a large value.

(B) *Fusion*   Since $S^{\circ,l} \approx S^{\circ,s}$, then it is not to be expected that there is a similar constancy in $\Delta_{fus}S_m^\circ$. Nevertheless, for metals, *Richard's rule* works quite well:

$$\Delta_{fus}S_m^\circ \approx 9 \text{ J mol}^{-1} \text{ K}^{-1} \tag{4.31}$$

which, if the transition temperature is 1000 K, gives $\Delta_{fus}H_m^\circ \approx 9$ kJ mol^{-1}, much smaller than for vaporization.

Semiconductors and inorganic compounds can have much larger values for $\Delta_{fus}S_m^\circ$ than for metals, so that Richard's rule should not be used in estimating this quantity for these materials.

(C) *Allotropic Transformations*   The enthalpies for the transformation between different polymorphic forms of both elements and compounds are important in the calculation of phase diagrams. These enthalpies are often referred to as the *relative lattice stabilities*. Some examples are shown in Figure 4.7. It can be seen that, at the beginning of a

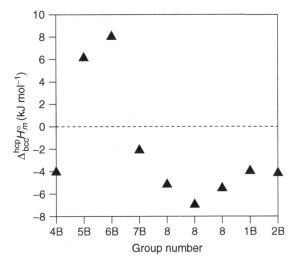

Figure 4.7 Relative lattice stability for some transition metals shown as the difference in the standard enthalpies between the bcc and hcp structures.

transition period, the hexagonal close-packed (hcp) structure is more stable than the bcc structure, but this trend is first reversed and then recovered towards the end of the period. Note that the values of $\Delta_{tr}H_m^\circ$ for solid/solid transformations are generally smaller than for fusion.

4.3 STABILITY AND CRITICAL PHENOMENA

In Chapter 2 we considered some basic EOS for gases and condensed phases. It is apparent from Figure 4.1, however, that the gas and liquid phases become indistinguishable at the critical point. As a result, there has been a desire to develop EOS which apply throughout the whole $p-V-T$ gas/liquid region, so that even isotherms both above and below the critical temperature can be represented analytically.

Figure 4.8 shows a $p-V$ diagram which includes two isotherms, one above and one below the critical temperature. Van der Waals first proposed an EOS which, although it has several shortcomings from a theoretical and practical standpoint, illustrates how both these super- and subcritical isotherms can be represented by one equation. The van der Waals equation is

$$p = \frac{RT}{V_m - b} - \frac{a}{V_m^2} \tag{4.32}$$

where a and b are parameters. The first term on the right-hand side is due to the attraction between the molecules and the second term accounts for repulsion forces between them. Being an analytical equation, it can be plotted for any

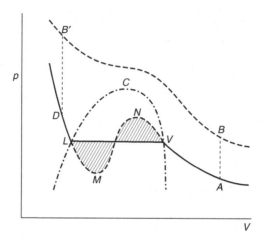

Figure 4.8 A loop in a subcritical isotherm when the van der Waals EOS is used. *Source*: I. Prigogine and R. Defay, *Chemical Thermodynamics*, Longman, 1954.

values of the parameters and variables. At a temperature below the critical, then, as can be seen in Figure 4.8, the shape of the isotherm is quite different from those found above the critical temperature. At very high temperatures, the van der Waals EOS gives isotherms which approach those for the perfect gas, while below T_c they show the so-called van der Waals loop.

The various parts of the continuous curve *AVNMLD* on the subcritical isotherm have distinct physical significances. It can be seen that for the part between N and M

$$\left(\frac{\partial p}{\partial V}\right)_T > 0 \qquad (4.33)$$

which means that such states are mechanically unstable (the volume must decrease with increase of pressure) and therefore not realizable in practice; that is, any point between N and M is unstable. The calculated isotherms have no physical meaning here and are just the result of an analytical representation which gives a continuous curve at all p and V_m. The loci of the points M and N, where

$$\left(\frac{\partial p}{\partial V}\right)_T = 0$$

at different temperatures gives the region marked unstable in Figure 4.9. The curve which separates the unstable region from the metastable region is called the *spinodal curve*. The portion of the curve marked VN in Figure 4.8 represents

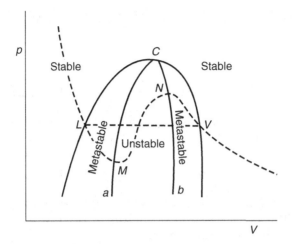

Figure 4.9 Stable, metastable, and unstable regions on unary phase diagram. The curve separating the metastable and unstable regions is called the spinodal, that between the stable and metastable regions the binodal. *Source*: I. Prigogine and R. Defag, *Chemical Thermodynamics*, Longman, 1954.

supersaturated vapor and the part LM corresponds with overexpanded liquid. These states might exist in metastable equilibrium up to the limit where the extensions enter the unstable region. Thus the region between the outer phase boundary and the inner envelope separating the unstable region can be marked metastable, as shown in Figure 4.9.

Note that, by following the path $ABB'D$ in Figure 4.8, it is possible to go from gas to liquid without at any stage observing the appearance of a new phase, in contrast to the situation if path $AVLD$ is followed.

4.4 GIBBS'S PHASE RULE

The phase rule is concerned with evaluating how many *field* variables can be fixed arbitrarily for a system consisting of several phases. We will meet this again when considering binary systems, but here we are concerned with only unary systems.

Associated with each phase in a one-component system are *three* field variables. These will usually be T, p, and μ.

If we have ϕ phases in the system, then there are 3ϕ field variables in total.

But these variables are not independent since, first, p, T, and μ_A are equal in all phases. If ϕ is the number of phases,

$$T^{(1)} = T^{(2)} \cdots = T^{(\phi)} \tag{4.34}$$

$$p^{(1)} = p^{(2)} \cdots = p^{(\phi)} \tag{4.35}$$

$$\mu_A^{(1)} = \mu_A^{(2)} \cdots = \mu_A^{(\phi)} \tag{4.36}$$

Second, we have a Gibbs–Duhem relation between the three field variables for each phase:

$$0 = -S \, dT + V \, dp + n_A \, d\mu_A$$

so that there are $3(\phi - 1) - \phi$ constraints placed on the total number of field variables.

The number of independent variables or, as it is more usually called in thermodynamics, the number of degrees of freedom F is, therefore

$$F = 3\phi - 3(\phi - 1) - \phi = 3 - \phi \qquad (4.37)$$

which is the usual way of writing Gibbs's phase rule for one-component systems. Figure 4.10 Illustrates the number of degrees of freedom for one-, two-, and three-phase equilibrium in a unary system.

In words, the rule states that the existence of a single phase requires the specification of two field variables (it has two degrees of freedom; the system is bivariant), the third field variable being fixed through the Gibbs–Duhem equation; when two phases coexist, specification of one field variable automatically fixes the other (there is one degree of freedom; the system is univariant); if three phases coexist in equilibrium the values of all the field variables are fixed (zero degrees of freedom; the system is invariant). Under no circumstances can four phases of a one-component system coexist in equilibrium.

It is interesting to note that the phase rule may not appear to apply at the critical point where there is also only a single phase present but where $F = 0$ rather than 2, as suggested by (4.37). The reason for this apparent breakdown is that the constraints given in (4.37) are no longer applicable since there is only one phase. Instead, some extra constraints are introduced. The same thing arises in binary systems and will be discussed in detail when encountered there.

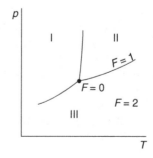

Figure 4.10 Number of degrees of freedom for one-, two-, and three-phase equilibrium in unary system.

EXERCISES

4.1 The following linearized equations give the approximate standard Gibbs energies of Zn in solid, liquid, and gaseous forms:

$$G^{\circ g}(\text{Zn,SER}, T) = 137{,}000 - 182.2T \quad \text{J/mol}$$

$$G^{\circ s}(\text{Zn,SER}, T) = 4175 - 53.95T \quad \text{J/mol}$$

$$G^{\circ l}(\text{Zn,SER}, T) = 11{,}200 - 64.1T \quad \text{J/mol}$$

Assume that the condensed-phase properties are independent of pressure and that the gas phase is perfect.

(a) Calculate the equilibrium vapor pressure in contact with both liquid and solid Zn at (i) 685 K and (ii) 700 K.

(b) Calculate which equilibrium has the lower G at these two temperatures.

(c) Does the stable state have a higher or lower vapor pressure than the metastable state?

(d) plot a $T - \mu_{\text{Zn}}$ diagram which shows these results by plotting the condensed-phase lines and vapor lines for fixed values of p_{Zn}.

(e) Is this field–field diagram a true phase diagram?

4.2 The following results were obtained for the melting point (mp) and the volume change accompanying the melting of a solid at different pressures:

p/MPa	mp/°C	$\Delta_{\text{fus}} V_m^*$/cm^3 mol^{-1}
0.1	97.6	2.787
200.0	114.2	2.362
400.0	129.8	2.072
600.0	142.5	1.873
800.0	154.8	1.711
1000.0	166.7	1.556
1200.0	177.2	1.398

Evaluate $\Delta_{\text{fus}} H^*$ as a function of pressure and plot the ratios of $\Delta_{\text{fus}} V_m^* / \Delta_{\text{fus}} V_m^{\circ}$ and $\Delta_{\text{fus}} H_m^* / \Delta_{\text{fus}} H_m^{\circ}$ as a function of pressure. Do these results support our MS&E assumption that $\Delta_{\text{fus}} H_m^* \approx \Delta_{\text{fus}} H_m^{\circ}$?

4.3 Derive (4.19) from (4.16).

4.4 Sketch a complete schematic $p-T$ phase diagram for Fe in the solid, liquid, and gaseous states. Use the information in Figures 4.1 and 4.2.

4.5 The vapor pressure of Si^l is represented by the equation

$$\log_{10} p \text{ (torr)} = 10.78 - 20,900 T^{-1} - 0.565 \log T$$

Calculate its enthalpies of vaporization at 1700 and 1500 K (undercooled liquid).

4.6 The vapor pressure of a pure element such as Cu, Si, and Ni is often expressed in either of the following equations:

$$\log_e p = A + \frac{B}{T} \quad \text{or} \quad \log_e p = A' + \frac{B'}{T} + C' \ln T$$

(a) Starting from the Clausius–Clapeyron equation, show under what thermodynamic condition do these two equations given above apply?

(b) Take the vapor pressures of Si tabulated as a function of temperature from a reference book such as the NIST-JANAF thermochemical tables and obtain values of A and B as well as A', B', and C' in the above two equations from 298.15 to their melting point. Compare the values calculated from either of these equations with the tabulated values.

4.7 The densities of α-Fe(bcc) and γ-Fe(fcc) are 7.57 and 7.63 g cm^{-3}, respectively, at 1184 K, 1 bar, when they are in equilibrium with each other. The enthalpies of transformation from α-Fe to γ-Fe are 900 J mol^{-1} at the transition temperature 1184 K and 1660 J mol^{-1} at 1100 K. Calculate the pressure under which both forms of Fe can coexist at 950 K. State the assumptions you make in order to carry out the calculation.

4.8 Starting from the equation $dG = V\,dp - S\,dT$, do a back-of-the-envelope calculation to estimate the pressure needed to transform α-Fe(bcc) to a closer packed γ-Fe(fcc) at room temperature given the entropy of α-Fe(bcc) to be about 55 J K^{-1} mol^{-1} and density to be about 7.5 g cm^{-3}.

4.9 The vapor pressure of liquid silver is given by

$$\log_{10}[p \text{ (bar)}] = 8.83 - \frac{14,400}{T} - 0.85 \log_{10} T$$

How much heat is required to evaporate 0.11 mol of Ag at 2300 K? The normal boiling point of silver is 2436 K.

4.10 Calculate the partial pressure of zinc over liquid zinc at 600°C using the thermodynamic data of Zn(l) and Zn(g) given below:

	800 K	900 K
gef(g)/J mol^{-1} K^{-1}	168.35	169.94
gef(l)/J mol^{-1} K^{-1}	52.672	55.865

$$\Delta_{sub}H^\circ(298\ \text{K})\ (\text{kJ mol}^{-1}) = 130.42$$

where $gef = -[G^\circ(T) - H^\circ(298\ \text{K})]/T$.

4.11 Calculated $\Delta_r G^\circ$ for the reduction of SiO_2(s) by Al(l) at 1200 K. Do you think that pure Al will reduce SiO_2 at this temperature?

5 Thermodynamics of Binary Solutions I: Basic Theory and Application to Gas Mixtures

Gas mixtures are a familiar type of solution phase. While gas-phase solutions are usually referred to as gas mixtures, this is not the case for condensed phases. There, the term *mixture* is usually used to refer to a mechanical mixture of the components and the term *solution* when one component actually dissolves in the other.

Water and alcohol form a continuous series of liquid solutions; we say they are completely *miscible*. Water and oil do not; they have no appreciable *mutual solubility*; they are nearly *immiscible*. Water and salt form a limited liquid solution. On continuing the addition of the salt, the solution becomes *saturated* with the solid phase.

Metals in the liquid and solid states also form solutions. In the latter case we speak of *solid solutions*. Solid solutions are further distinguished as being either *substitutional* or *interstitial*. In substitutional solid solutions all the atoms are on the one lattice. Familiar examples are (Cu,Zn), (Cu,Sn), (Cu,Ni), (Si,Ge) alloys. In interstitial solid solutions the smaller atoms reside in the interstices of the sublattice formed by the larger atoms. It is convenient to adopt the notation (M):(I) to represent the occupation of the metal and interstitial sublattices. Familiar examples using this notation are (Fe):(C) and (Pd):(H).

5.1 EXPRESSING COMPOSITION

Three methods are commonly used:

(i) Mole fraction or percent
(ii) Mass fraction or percent
(iii) Volume fraction or percent

Volume percent is the usual way of expressing the composition of gas mixtures. Because gas volumes change markedly with p and T, it is necessary to specify

Materials Thermodynamics. By Y. Austin Chang and W. Alan Oates
Copyright 2010 John Wiley & Sons, Inc.

both. The most common way is to refer the volumes to STP (standard temperature and pressure), that is, $T = 273.15$ K and $p = 1$ bar.

Mole and mass fractions or percents are the most commonly used ways of expressing the composition of condensed phases. The definitions of mole fraction and weight percent require that

$$\sum_i x_i = 1 \qquad \sum_i \text{wt} \% i = 100 \tag{5.1}$$

Consider a liquid or substitutional solid solution such as (Cu,Zn). The term mole fraction refers to 1 mol of alloy atoms, that is, to $Cu_{1-x}Zn_x$, where the subscript x refers to the mole fraction of Zn and $1 - x$ to the mole fraction of Cu:

$$x_{Zn} = \frac{n_{Zn}}{n_{Zn} + n_{Cu}} \tag{5.2}$$

$$\text{wt} \% \, Zn = \frac{m_{Zn}}{m_{Zn} + m_{Cu}} \times 100 \tag{5.3}$$

Conversion from one composition variable to the other is straightforward since

$$n_{Zn} = \frac{m_{Zn}}{A_{r,Zn}} \tag{5.4}$$

For a solution of binary compounds such as FeO–MnO the mole fraction is

$$x_{MnO} = \frac{n_{MnO}}{n_{MnO} + n_{FeO}} \tag{5.5}$$

and equivalently for the weight percent. Conversion between composition variables in this case is obtained by using

$$n_{MnO} = \frac{m_{MnO}}{M_{r,MnO}} \tag{5.6}$$

5.2   TOTAL (INTEGRAL) AND PARTIAL MOLAR QUANTITIES

The magnitudes of the extensive thermodynamic properties of a solution phase are rarely found to be the composition-weighted average of the properties of the pure components, that is, that of the mechanical mixture. For example, for the volume of a solution of water and alcohol:

100 ml water + 100 ml alcohol ≠ 200 ml of solution but ≈ 190 ml of solution

In this instance, there is a volume contraction (at the microscopic level, there is strong attraction between the molecules of water and alcohol, resulting in a reduction of the solution volume).

Similar nonlinear behavior is found in all types of solutions for the other extensive thermodynamic quantities, U, H, G, A, S, ...

Any extensive thermodynamic quantity Y, at constant p and T, is a function of the amounts of substance of the components, A and B, that is,

$$Y|_{p,T} = Y(n_A, n_B) \qquad (5.7)$$

and the differential quantity is given by

$$dY|_{p,T} = \sum_i Y_i\, dn_i \qquad (5.8)$$

where Y_i is the partial derivative at constant p and T and n_j is the amount of substance of the other component(s):

$$\boxed{Y_i = \left(\frac{\partial Y}{\partial n_i}\right)_{p,T,n_j}} \qquad (5.9)$$

This last equation defines a *partial molar quantity*. Note that these partial derivatives are only referred to as partial molar quantities when the derivatives refer to constant p and T.

Extensive thermodynamic functions are homogeneous functions of the first degree, that is,

$$kY(n_i) = Y(kn_i) \qquad (5.10)$$

where k is a constant, and by Euler's theorem for such functions, the integral quantity is related to the partial molar quantities by

$$\boxed{Y = \sum_i n_i Y_i} \qquad (5.11)$$

The difference between a total (or integral) property and a partial molar property is clear from the two experiments sketched in Figure 5.1. The sketch on the left shows how the total or integral enthalpy of mixing $\Delta_{\text{mix}} H$ for a liquid (A,B) alloy at a particular composition, $A_{0.7}B_{0.3}(l)$, at constant temperature and pressure. The relative amounts of the pure components required to give the desired composition are simply mixed. The sketch on the right shows how the partial molar enthalpy of mixing of A, ΔH_A, in a liquid alloy of the same composition, $A_{0.7}B_{0.3}(l)$, and at the same temperature and pressure is measured. A very small amount of component A is added to a relatively large amount of the alloy. The result from this experiment is independent of the size of the alloy bath as long as it is very large compared with the amount of A added.

Because of the relation between total (extensive) and partial molar quantities, $Y = \sum_i n_i Y_i$, we can see that all the relations previously met for the extensive

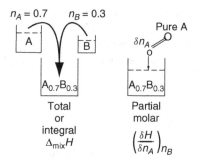

Figure 5.1 Difference between total or integral thermodynamic property and partial molar property. Both mixing processes are carried out at constant p and T.

properties of unary systems apply equally well to the partial molar quantities in solutions (remember that the partial molar quantities are only defined at constant p and T). Some examples are

$$G_i = H_i - T S_i \tag{5.12}$$

$$H_i = U_i + p V_i \tag{5.13}$$

$$\Delta_{mix} G_i = \Delta_{mix} H_i - T \Delta_{mix} S_i \tag{5.14}$$

$$\left(\frac{\partial G_i}{\partial T} \right)_p = -S_i \tag{5.15}$$

$$\left(\frac{\partial G_i}{\partial p} \right)_T = -V_i \tag{5.16}$$

5.2.1   Relations between Partial and Integral Quantities

We will show how (5.9) and (5.11) can be used to derive some important relations between partial and integral quantities.

The molar form of (5.8) can be obtained for a binary solution by dividing by $n_A + n_B$:

$$dY_m = Y_A \, dx_A + Y_B \, dx_B \tag{5.17}$$

and since $x_A + x_B = 1$, then $dx_A = -dx_B$, giving

$$dY_m = (Y_B - Y_A) \, dx_B \tag{5.18}$$

or

$$\frac{dY_m}{dx_B} = Y_B - Y_A \tag{5.19}$$

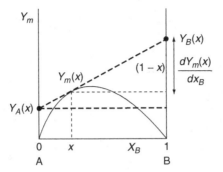

Figure 5.2 Tangent–intercept relation between partial and integral molar quantities at constant p and T.

Graphically, this means that, for a binary solution, the tangent slope at a point on a plot of an integral molar quantity is equal to the difference in the partial molar properties at that point.

Similarly, the molar form of (5.7) for a binary solution is given by

$$Y_m = x_A Y_A + x_B Y_B = (1 - x_B)Y_A + x_B Y_B \qquad (5.20)$$

Combination of (5.19) and (5.20) yields equations from which the individual partial molar quantities can be obtained from the integral molar quantity:

$$\boxed{Y_A = Y_m - x_B \frac{dY_m}{dx_B}} \qquad (5.21)$$

$$\boxed{Y_B = Y_m + (1 - x_B)\frac{dY_m}{dx_B}} \qquad (5.22)$$

These equations also have a very simple graphical interpretation which is shown in Figure 5.2. The partial molar quantities are given by the intercepts on the ordinate (property) axes of the tangent to the integral property curve at the composition of interest.

Example 5.1 Calculation of Partial Molar from Integral Enthalpies of Mixing
The enthalpy of mixing of Cu(l) with Sn(l) to form 1 mol of liquid $Cu_{1-x}Sn_x$ alloy as a function of composition at 1400 K (and $p = 1$ bar) is shown in Figure 5.3. The tangent shown refers to a composition $x_{Sn} = 0.2$. The two tangent intercepts at this composition are shown on the two ordinate axes. These two values for the partial molar properties are then shown by dots in Figure 5.4 for this particular composition. The curves in the latter figure are obtained by repeating this procedure at all compositions. The advantage of having an analytical representation of the integral property and then being able to do the necessary differentiations analytically is obvious. We will return to this point later.

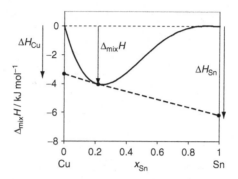

Figure 5.3 $\Delta_{\mathrm{mix}} H_m$ for liquid $Cu_{1-x}Sn_x$ alloys at 1400 K and 1 bar. The reference states are Cu(l) and Sn(l).

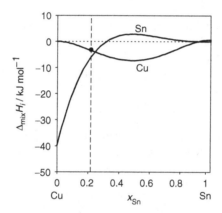

Figure 5.4 Partial molar enthalpies of mixing for liquid $Cu_{1-x}Sn_x$ alloys at 1400 K and 1 bar. The reference states are Cu(l) and Sn(l).

5.2.2 Relation between Partial Quantities: the Gibbs–Duhem Equation

The fourth Gibbs equation for a solution phase is

$$dG = -S\,dT + V\,dp + \sum_i \mu_i\,dn_i \tag{5.23}$$

Since one of the independent variables, n_i, is extensive, while the other two, p and T, are independent of the size of the system, integration of this equation gives the equation

$$G = \sum_i n_i \mu_i \tag{5.24}$$

where the μ_i are functions of p and T.

The result from this integration also follows from Euler's equation for homogeneous functions of degree 1 (see Appendix B):

The total differential of (5.24) is

$$dG = \sum_i n_i \, d\mu_i + \sum_i \mu_i \, dn_i \tag{5.25}$$

Subtracting (5.23) from (5.24) gives

$$0 = -S \, dT + V \, dp + \sum_i n_i \, d\mu_i \tag{5.26}$$

which is the Gibbs–Duhem equation for a solution phase.

At constant p and T it becomes

$$0 = \sum_i n_i \, d\mu_i \tag{5.27}$$

from which we see that the chemical potentials in a solution phase are not independent.

The molar form of (5.27) for a binary system is obtained by dividing by $n_A + n_B$ and taking the derivative with respect to x_B:

$$x_A \left(\frac{\partial \mu_A}{\partial x_B} \right)_{p,T} + x_B \left(\frac{\partial \mu_B}{\partial x_B} \right)_{p,T} = 0 \tag{5.28}$$

which, on integrating, gives

$$\int d\mu_B = - \int \frac{1 - x_B}{x_B} \left(\frac{\partial \mu_A}{\partial x_B} \right)_{p,T} dx_B \tag{5.29}$$

We can see from this equation that, if the composition variation of one chemical potential is known in a binary solution for a given p and T, then the other chemical potential can be evaluated. This represents another example of using thermodynamics to avoid, in principle, having to carry out extraneous measurements.

The constant p and T version of the Gibbs–Duhem equation can be equally applied to the other partial molar properties of a solution:

$$0 = \sum_i n_i \, dH_i \tag{5.30}$$

5.3   APPLICATION TO GAS MIXTURES

5.3.1   Partial Pressures

Pressure is a field quantity (see Chapter 4). In a gas mixture we wish to assign the separate contributions of the components to the total pressure. Partial pressures are defined to do just this; that is, their sum is equal to the total pressure:

$$\boxed{p_i = x_i p} \qquad (5.31)$$

from which we see that the definition ensures that the sum of the partial pressures equals the total pressure:

$$\sum_i p_i = \sum_i x_i p = p \sum_i x_i = p \qquad (5.32)$$

Because it is a definition, this equation applies independently of whether the gas is perfect or not. Note that (5.31) is *not*, as is sometimes claimed, Dalton's so-called law of partial pressures. Dalton found experimentally that the total pressure of a gas mixture, in a constant-volume enclosure at constant T, was *approximately* equal to the sum of the pure component pressures in the same enclosure: $p \approx \sum_i p_i^*$

5.3.2 Chemical Potentials in Perfect Gas Mixtures

We have already seen in (2.28) in Chapter 2 that, for a unary gas,

$$G_m^*(T, p) - G_m^\circ(T, p^\circ) = RT \ \log_e \left(\frac{p^*}{p^\circ} \right) \qquad \text{perfect gas} \qquad (5.33)$$

We can go through a derivation similar to that used for obtaining (5.33) for a component of a perfect gas mixture at constant T:

$$dG_i|_T = V_i \, dp_i \qquad (5.34)$$

and for a perfect gas

$$dG_i|_T = \frac{RT}{p_i} \, dp_i \qquad (5.35)$$

On integration this gives, with respect to the standard conditions,

$$G_i(T, p^*) - G_i^\circ(T, p_i^\circ) = RT \ \log_e \left(\frac{p_i}{p_i^\circ} \right) \qquad (5.36)$$

or, for any conditions,

$$\boxed{\mu_i(T, p_i) - \mu_i^\circ(T) = RT \ \log_e \left(\frac{p_i}{p_i^\circ} \right)} \qquad \begin{array}{l} \text{component} \\ \text{perfect gas mixture} \end{array} \qquad (5.37)$$

This equation is used very often in MS&E calculations, where it is usually safe to assume that gases are perfect. It gives us a recipe for calculating μ_i for a component of a gas mixture from observable quantities (gas analysis).

5.3.3   Real Gas Mixtures: Component Fugacities and Activities

Defining the fugacity of a component in a mixture is similar to that already met with in Chapter 2 for a pure gas—for a real gas we keep the same form as (5.37) in order to define the fugacity of a component:

$$\mu_i(T, p_i) - \mu_i^{\circ}(T) = RT \, \log_e \left(\frac{f_i}{f_i^{\circ}} \right) \quad \text{real gas mixture} \quad (5.38)$$

where f_i° is the standard state fugacity of component i at p_i°, T.

The activity of a component is *defined* as the ratio of its fugacity in the state it happens to be in to its fugacity in the standard state:

$$\boxed{a_i = \frac{f_i}{f_i^{\circ}}} \quad (5.39)$$

When the gas phase can be assumed perfect, then the activity of a component can be obtained from

$$a_i = \frac{p_i}{p_i^{\circ}} \quad (5.40)$$

The relation between chemical potential and activity is given by

$$\boxed{\mu_i(T, p_i) - \mu_i^{\circ}(T) = RT \, \log_e a_i} \quad (5.41)$$

and we see that the functions fugacity and activity are just alternative ways of expressing the chemical potential of a component in a gas mixture. The same functions are used for components of solid and liquid solutions.

EXERCISES

5.1   The integral enthalpy of mixing of liquid (Cu,Zn) solution can be approximated by the following equation using Cu(l) and Zn(l) as the reference states: $\Delta_{\text{mix}} H_m = -19,250 x_{\text{Cu}} x_{\text{Zn}}$ J mol^{-1} at 1400 K.

(a)   Derive the corresponding partial quantities of Cu and Zn, respectively.

(b)   Show the two partial equations are internally consistent with the Gibbs–Duhem equation.

(c)   Plot the integral enthalpy values at 1400 K as a function of composition. Draw a tangent at 50 mol% Zn and obtain the two partial quantities graphically as shown in Figure 5.3.

(d) Plot the variation of the integral and partial quantities graphically and make a comparison with the data shown in Figures 5.3 and 5.4. Please make any comment you may have on the different behaviors between the two alloy systems.

5.2 (a) Calculate the enthalpy effect ΔH for dissolving 1 mol of solid nickel in 9 mol of liquid copper at 1473 K, assuming an ideal solution is formed. In other words, the enthalpy of mixing liquid (Cu,Ni) alloys is negligibly small and can be taken to be zero. The enthalpies of melting for Ni and Cu are 17,470 J mol^{-1} at $T_{\text{fus}} = 1726$ K and 13,055 J mol^{-1} at $T_{\text{fus}} = 1356.6$ K.

(b) Calculate the entropy change of the system and of the surroundings for this process.

5.3 (a) A stream of argon gas is passed in a closed system over a boat containing mercury at 273 K. The flow rate of the argon is slow enough to allow this gas to become saturated with mercury vapor. The total volume of nitrogen used is 22 liters measured at 293.15 K and 1 bar. The argon was found to contain 0.0674 g of mercury. Calculate the vapor pressure of mercury at 273.15 K.

(b) The same experiment is carried out with a sodium amalgam (Na + Hg) in which the atomic fraction of sodium is 0.122; 22 liters of argon gas saturated with mercury is found to contain 0.0471 g of mercury. Take pure mercury as the reference state and calculate the chemical potential of Hg relative to that of Hg(l), that is, $\mu_{\text{Hg}} - \mu_{\text{Hg}}^{\circ}$.

5.4 Copper (Cu) and nickel (Ni) form a continuous series of solutions in both the liquid- and solid state. The crystal structure of both solid elements is face-centered cubic.

The following experiments have been carried out to measure the enthalpy of solution of Cu(l, 1823 K, 1 bar) and that of Ni(l, 1823 K, 1 bar) in a liquid Cu$_{0.6}$Ni$_{0.4}$ alloy, also held at 1823 K, 1 bar.

Specifically a sample of 0.2 g of liquid Cu held at 1823 K, 1 bar, is dropped into a large bath of liquid Cu$_{0.6}$Ni$_{0.4}$ alloy (say, 1000 g) also held at 1823 K and 1 bar. The enthalpy change measured is 6.14 J (endothermic or heat flow into system).

A similar experiment is also carried out by dropping 0.2 g of Ni held at 1823 K and 1 bar into the same bath under identical conditions. The enthalpy change measured is 14.96 J (also endothermic).

Element	A_r	$S_m^{\circ,\text{s}}(298 \text{ K})$ /J K^{-1} mol^{-1}	$\langle C_p^{\circ,\text{s}} \rangle$ /J K^{-1} mol^{-1}	$\langle C_p^{\circ,\text{l}} \rangle$ /J K^{-1} mol^{-1}	T_{fus}° / K	$\Delta_{\text{fus}} H_m^{\circ}(T_{\text{fus}}^{\circ})$ /J	Vapor Pressure/bar
Cu	63.54	28.5	27.2	32.6	1356.6	13055.0	6.80×10^{-4}
Ni	58.71	33.5	32.5	43.1	1726	17470.0	2.76×10^{-5}

Use these experimental data as well as any other thermodynamic data given in the above table (note that you may not need all the data in the table) to calculate the following (stating any assumptions you may have to make in carrying out some of the calculations):

(a) Relative partial molar enthalpy of Cu, $\Delta_{mix} H_{Cu}(1823$ K, 1 bar), in $Cu_{0.6}Ni_{0.4}$

(b) Relative partial molar enthalpy of Ni, $\Delta_{mix} H_{Ni}(1823$ K, 1 bar), in $Cu_{0.6}Ni_{0.4}$

(c) Integral molar enthalpy of mixing, $\Delta_{mix} H_m$, for $Cu_{0.6}Ni_{0.4}$ (l, 1823 K, 1 bar)

(d) Relative partial molar enthalpy of Cu at 1823 K and 1 bar but with respect to Cu(s)

(e) Relative partial molar enthalpy of Cu at 1823 K and 1 bar but with respect to the SER reference state

(f) Equilibrium partial pressures of Cu and Ni over the alloy $Cu_{0.6}Ni_{0.4}$ at 1823 K. You may assume that $\Delta_{mix} S_m$ is the same as that for an ideal solution.

6 Thermodynamics of Binary Solutions II: Theory and Experimental Methods

6.1 IDEAL SOLUTIONS

Just as the perfect gas mixture provides a useful concept when discussing the properties of a real gas mixture, it is similarly useful to define ideal solutions for condensed phases and then discuss the behavior of real solutions in terms of their deviation from the defined ideal behavior.

With gases, it was clear how to define a perfect gas—one whose EOS is given by $pV = nRT$, which results when intermolecular interactions are negligible. Real gases at moderate pressures, and particularly at high temperatures, approach this defined ideal behavior. In the case of liquid and solid solutions, however, we have the situation where the balance between attractive and repulsive forces between the atoms or molecules is actually responsible for the stability of the condensed phase, so that we cannot define a condensed-phase ideal solution on the basis of zero atomic or molecular interactions. Such a definition is not necessary, however, since we are only interested in property changes on mixing to form a solution and we might, therefore, define an ideal solution as one where there are negligible effects brought about by changes in the atomic or molecular interactions on forming the solution. On the basis of this definition, we only expect a real solution to approach this ideal behavior when the components are chemically identical. Mixtures or solutions of isotopes are, therefore, the most likely candidates for forming an ideal solution. An alloy of the chemically similar elements Ag and Au might be expected to approach ideal solution behavior while chemically dissimilar elements like Al and Ti would not.

For a component i of a perfect gas mixture the chemical potential relative to that for the pure component at pressure p_i^* is given by

$$\mu_i(p_i, T) - \mu_i^*(p_i^*, T) = RT \log_e \left(\frac{p_i}{p_i^*} \right) \tag{6.1}$$

Materials Thermodynamics. By Y. Austin Chang and W. Alan Oates
Copyright 2010 John Wiley & Sons, Inc.

and from the definition of partial pressure:

$$p_i = x_i p_i^* \qquad (6.2)$$

In such a perfect gas mixture we may write the chemical potential of the component i in terms of molar concentration as

$$\mu_i(p_i, T) - \mu_i^*(p_i^*, T) = RT \log_e \left(\frac{x_i p_i^*}{p_i^*} \right) \qquad (6.3)$$

$$\mu_i(p_i, T) - \mu_i^*(p_i^*(x_i = 1), T) = RT \log_e x_i \qquad (6.4)$$

In liquid or solid solutions we are interested in solutions at, effectively, $p = 1$ bar (in MS&E, the effect of pressure on condensed-phase properties is generally negligible). Comparison with (6.3) leads to a plausible definition of an ideal solution for condensed phases:

$$\mu_i(x_i, T) - \mu_i^*(x_i = 1, T) \approx \mu_i(x_i, T) - \mu_i^{\circ}(x_i = 1, T)$$
$$= RT \log_e x_i \qquad (6.5)$$

which we will usually write as

$$\boxed{\Delta_{\text{mix}}\mu_i^{\text{id}} = RT \log_e x_i} \qquad (6.6)$$

Recall that the subscript mix is used whenever the reference states for the pure components have the same structure as that of the phase of interest.

It should be made clear that an ideal solution could be defined differently from (6.6); the definition is in the hands of the thermodynamicist and is not an absolute. The aim is to use a definition such that the properties of the defined ideal solution are approached by real solutions.

Solutions conforming to the definition given in (6.6) are said to be Raoultian ideal. But other definitions of ideal solutions, more appropriate for other types of solution, are used. For the moment, it should be emphasized that we have defined *an* ideal solution and not *the* ideal solution. This particular definition is useful when discussing liquid metal alloys and solid substitutional alloys. For the remainder of this chapter, we will usually drop the prefix Raoultian and simply use the term ideal solution.

The following equations for ΔH_i^{id}, ΔS_i^{id}, and ΔV_i^{id}, at constant x_i, for our defined ideal solution are readily obtained:

$$\Delta_{\text{mix}} H_i^{\text{id}} = \frac{\partial(\Delta\mu_i^{\text{id}}/T)}{\partial(1/T)} = \frac{\partial(R \ln x_i)}{\partial(1/T)} = 0 \qquad (6.7)$$

$$\Delta_{\text{mix}} S_i^{\text{id}} = -\frac{\partial \Delta\mu_i^{\text{id}}}{\partial T} = -\frac{\partial(RT \ln x_i)}{\partial T} = -R \log_e x_i \qquad (6.8)$$

$$\Delta_{\text{mix}} V_i^{\text{id}} = \frac{\partial \Delta \mu_i^{\text{id}}}{\partial p} = \frac{\partial (RT \log_e x_i)}{\partial p} = 0 \tag{6.9}$$

The integral molar quantities may be obtained from these partial molar quantities by using, for any property Y, $Y_m = \sum_i x_i Y_i$. For a binary solution (A,B)

$$\Delta_{\text{mix}} G_m^{\text{id}} = RT[x_A \log_e x_A + x_B \log_e x_B] \tag{6.10}$$

$$\Delta_{\text{mix}} H_m^{\text{id}} = 0 \tag{6.11}$$

$$\Delta_{\text{mix}} S_m^{\text{id}} = -R[x_A \log_e x_A + x_B \log_e x_B] \tag{6.12}$$

$$\Delta_{\text{mix}} V_m^{\text{id}} = 0 \tag{6.13}$$

The compositional variations of the partial and integral entropies and Gibbs energies (in dimensionless units) are shown in Figures 6.1 and 6.2 for a Raoultian ideal solution in a binary system. The integral and partial enthalpies and volumes of mixing are zero.

As shown in Figure 6.2, the Gibbs energy of mixing for this ideal solution is symmetrical with respect to composition; it attains a minimum value of $-0.69315R$ at the equi-atomic composition and approaches zero at the pure component elements. On the other hand, the partial quantities change much more markedly with composition due to the logarithmic term involved:

$$x_i \to 1 \qquad \Delta S_i, \, \Delta G_i \to 0$$
$$x_i \to 0 \qquad \Delta S_i \to +\infty \qquad \Delta G_i \to -\infty$$

At the midcomposition, the integral and partial Gibbs energies attain the same value in the defined ideal solution since the horizontal tangent to the integral Gibbs energy curve $\Delta G / RT$ at this point intercepts the ordinate axes at the same value as that of the integral quantity. These intercepts are the chemical potentials (or partial molar Gibbs energies in this case) of the component elements for the alloy at the equi-atomic composition.

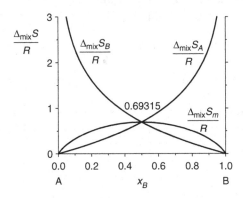

Figure 6.1 Integral and partial molar entropies for Raoultian ideal solution as defined in (6.6).

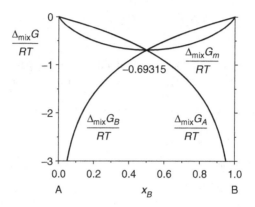

Figure 6.2 Integral and partial molar Gibbs energies for Raoultian ideal solution as defined in (6.6).

6.1.1 Real Solutions

The experimental results for many real systems indicate deviations from ideal solution behavior and it is useful to define quantities which are suited to describing their behavior. There are several commonly used ways of doing this. One of these, already met in Chapter 5, is to retain the form of (6.5) but to replace the mole fraction with another dimensionless quantity called the *activity*:

$$\mu_i(x_i, T) - \mu_i^\circ(x_i = 1, T) = RT \log_e a_i \qquad (6.14)$$

Comparison of (6.14) with (6.6) shows that, for an ideal solution,

$$\boxed{a_i(x_i) = x_i} \quad \text{Raoultian ideal, } p^\circ = 1 \text{ bar} \qquad (6.15)$$

The variation of component activities with composition for a Raoultian ideal solution is shown in Figure 6.3. Equation (6.14) can be rewritten as

$$\mu_i(x_i, T) - \mu_i^\circ(x_i = 1, T) = RT \log_e x_i + RT \log_e \gamma_i \qquad (6.16)$$

where γ_i is known as the activity coefficient or, more precisely, as the Raoultian activity coefficient.

Note that, because of the way that activity is defined, a solution at a pressure other than 1 bar in which the compositional variation of the chemical potential is given by

$$d\mu_i|_{p,T} = RT \, d \log_e x_i \qquad (6.17)$$

does not obey (6.15).

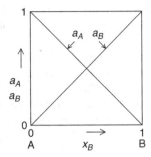

Figure 6.3 Variation of component activities with composition for a Raoultian ideal solution.

Another way for describing the properties of the real solutions is to use *departure* or *excess* functions. These also describe the deviation from the defined ideal solution behavior. The excess integral and partial molar properties are defined by

$$Y^{\mathrm{E}} = \Delta_{\mathrm{mix}} Y - \Delta_{\mathrm{mix}} Y^{\mathrm{id}} \tag{6.18}$$

$$Y_i^{\mathrm{E}} = \Delta_{\mathrm{mix}} Y_i - \Delta_{\mathrm{mix}} Y_i^{\mathrm{id}} \tag{6.19}$$

As will become apparent later, there are definite advantages in defining such new functions. The most important excess properties are

$$G_m^{\mathrm{E}} = \Delta_{\mathrm{mix}} G_m - \Delta_{\mathrm{mix}} G_m^{\mathrm{id}} \tag{6.20}$$

$$H_m^{\mathrm{E}} = \Delta_{\mathrm{mix}} H_m - \Delta_{\mathrm{mix}} H_m^{\mathrm{id}} = \Delta_{\mathrm{mix}} H_m \tag{6.21}$$

$$\mu_i^{\mathrm{E}} = \Delta_{\mathrm{mix}} \mu_i - \Delta_{\mathrm{mix}} \mu_i^{\mathrm{id}} \tag{6.22}$$

It can be seen that the excess chemical potential is related to the activity coefficient by

$$\mu_i^{\mathrm{E}} = RT \, \log_e \left(\frac{a_i}{x_i} \right) = RT \, \log_e \gamma_i \tag{6.23}$$

Note that, while $\Delta_{\mathrm{mix}} \mu_i = \Delta_{\mathrm{mix}} G_i \to -\infty$ as $x_i \to 0$, μ_i^{E} and $\Delta_{\mathrm{mix}} H_i$ remain finite in this limit, since they do not contain the logarithmic term associated with the mixing entropy.

6.1.2   Dilute Solution Reference States

So far we have concentrated on using the pure substance reference state. Sometimes the required pure substance may not exist, as, for example, when only dilute solutions can be investigated experimentally. In such cases, it is more convenient and perhaps necessary to select a more appropriate reference state for the solute (the pure substance reference state can be used for the solvent). Several are possible and we illustrate with just the one—the mole fraction infinitely dilute solution reference state.

In the pure substance reference state $\gamma_i \rightarrow 1$ as $x_i \rightarrow 1$. When the mole fraction infinitely dilute reference state is chosen, a new activity coefficient, f_i, is defined so that $f_i \rightarrow x_i$ as $x_i \rightarrow 0$. The differences between the pure substance and mole fraction infinitely dilute solution reference states, assuming results are available over the whole concentration range, are shown in Figure 6.4. The differences are also apparent in the definitions:

$$\mu_i = \mu_i^\circ + RT \log_e x_i + RT \log_e \gamma_i \tag{6.24}$$

$$= \mu_i^\ominus + RT \log_e x_i + RT \log_e f_i \tag{6.25}$$

where we have used $\mu_i^\ominus$ for the standard reference state for the mole fraction infinitely dilute solution to distinguish it from the pure substance standard state. As can be seen from (6.25), the value of $\mu_i^\ominus$ can be obtained from dilute solution results by plotting $\mu_i - RT \log_e x_i$ versus x_i. The intercept at $x_i = 0$ is then the value of $\mu_i^\ominus$, since this is where $f_i = 1$. If results are available over the whole concentration range, comparison of (6.24) with (6.25) gives the relation between f_i and γ_i:

$$f_i = \gamma_i \exp\left(\frac{\mu_i^\circ - \mu_i^\ominus}{RT}\right) \tag{6.26}$$

It is sometimes found experimentally that f_i appears to remain approximately constant at $f_i = 1$ (or, alternatively, γ_i is constant, at a value usually represented as γ_i°) in very dilute solution of the solute i. Under these circumstances, the linear relation a_i(pure substance scale) $= \gamma_i^\circ x_i$ is known as *Henry's law*. It is more often used in dilute aqueous solutions than in alloy solutions.

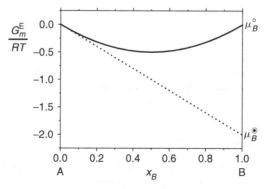

Figure 6.4 Difference between pure substance and infinitely dilute solution reference states.

In the dilute solution region, it is possible to use an interpolation formula as an approximation for calculating the activity coefficient of a solute in dilute concentration in a multicomponent solution. The values for the individual solutes are stored as *interaction coefficients* in binary and ternary systems:

$$\epsilon_i^i = \left(\frac{\partial \log_e f_i}{\partial x_i} \right) \qquad \epsilon_i^j = \left(\frac{\partial \log_e f_i}{\partial x_j} \right)$$

Then, in the multicomponent solution,

$$\log_e f_i = 1 + x_i \epsilon_i^i + \sum_j x_j \epsilon_i^j \qquad (6.27)$$

It is emphasized that such a linear interpolation should only be used for solutes in very dilute solutions.

6.2 EXPERIMENTAL METHODS

As we saw in Chapter 5, it is possible, in principle, to obtain all the *mixing* thermodynamic quantities, that is, both the integral and the partial Gibbs energies, enthalpies, and entropies of mixing for a *binary* solution phase when the chemical potential of *one* component is measured as a function of composition and temperature. The Gibbs–Duhem equation provides the means for doing this.

In practice, however, the values for the derived mixing entropies and enthalpies obtained by this procedure may not always be reliable, even though the Gibbs energy values themselves may be satisfactory (there are practical difficulties in obtaining accurate slopes and intercepts from a plot of $\log_e a_i$ versus reciprocal temperature).

Improved approaches to obtaining accurate thermodynamic properties of solutions are to:

(a) Determine simultaneously the chemical potentials of *both* components. This can often be done by mass spectrometry measurements of the component vapor pressures.

(b) Determine the chemical potential of one component, for example, by vapor pressure measurements, together with either the integral or partial enthalpy of mixing using calorimetry.

The methods which are used in obtaining the high-temperature thermodynamic properties of alloys can be summarized as follows:

1. First-principles calculations of the 0 K total energies plus model calculations of excitational energies
2. *Substance* calorimetry

   (a) Heat capacity, C_p
   (b) Enthalpy difference, $H(T_2) - H(T_1)$
3. *Reaction* calorimetry
   (a) Integral enthalpies of mixing, $\Delta_{mix} H_m$
   (b) Direct
   (c) Solution
   (d) Partial molar enthalpies of mixing, $\Delta_{mix} H_i$
4. Chemical potential measurements, μ_i, via one of the following:
   (a) Phase equilibria
   (b) Chemical equilibria
   (c) Electrochemical equilibria

Substance and reaction calorimetry for unary systems was discussed in Chapter 2. There are no differences in the methods used for obtaining alloy formation properties from those used for compound formation properties and so it is unnecessary to consider them again here.

The principle behind the measurement of a component chemical potential in a phase is, most commonly, to equilibrate that phase with another phase in which its chemical potential is already known (phase equilibria). We anticipate the important relation, to be discussed in Chapter 8, that, for equilibrium between two phases, the chemical potentials of any given mobile component are equal in the two phases. If chemical reactions are involved between the phases involved, then a modified equilibrium criterion, to be discussed in detail in Chapter 18, is used. Similarly, when electrochemical reactions are involved between the phases (electrochemical equilibria), another equilibrium relation is necessary.

6.2.1 Chemical Potential Measurements

6.2.1.1 Phase Equilibria. The relation used is $\mu_i' = \mu_i''$, where the value in one phase is known. The latter is usually the gas phase at low pressures, where $\mu_i = \mu_i^\circ + RT \log_e(p_i/p_i^*)$ can be used with high accuracy.

Most of the older vapor pressure methods measure the total pressure rather than the individual component vapor pressures. As a result, these methods are suitable only for alloys where the vapor pressures of the two component elements differ from each other by at least three orders of magnitude. The measurement of vapor pressures at high temperature is not straightforward. Either the whole apparatus must be at a high temperature to prevent any undesirable condensation or some pressure-balancing device must be introduced to isolate the high-temperature reactive system from a room temperature inert gas pressure-measuring system.

An example is provided by liquid (Pb,Sn) alloys. For this binary system, the vapor pressure of Pb(l) is many orders of magnitude higher than that of Sn(l). Accordingly, the measured total pressures over (Pb,Sn) alloys are essentially the partial pressures of Pb, with the partial pressure of Sn being negligibly small if the concentration of Pb is at all significant. When the partial pressure of Pb is measured at a constant temperature as a function of composition from pure Pb to nearly pure Sn, the chemical potential or activity of Pb can be calculated from

$$a_{Pb} = \frac{p_{Pb}}{p_{Pb}^*} \approx \frac{p_{tot}}{p_{Pb}^*} \tag{6.28}$$

The activity of Sn can then be obtained from those for Pb via the Gibbs–Duhem relationship and the integral Gibbs energy from the two partial quantities.

A relatively new method by which it is possible to simultaneously determine the vapor pressures and therefore the activities of both components is known as KEMS (Knudsen effusion mass spectrometry). The simultaneous mass spectrometric measurements can, however, only be carried out when the temperatures and alloy components and compositions are such that the component vapor pressures do not differ by too many orders of magnitude.

6.2.1.2   *Chemical Equilibria.* To illustrate, we present the example of using a mixture of CO(g) and CO_2(g) gases to obtain the activity of C in ferrous alloys. Graphite, C(gr), is used as the reference state for C. The following chemical reaction represents the chemical equilibrium between a mixture of CO and CO_2 and C in an Fe alloy:

$$CO_2(g) + C(gr) = 2CO(g)$$

The equilibrium constant K_p for this reaction is

$$K_p = \frac{p_{CO}^2}{p_{CO_2} a_{C(gr)}} \tag{6.29}$$

The value of K_p is obtained from the value of the p_{CO}^2 / p_{CO_2} ratio in equilibrium with pure C(gr) so that the activity of C(gr) in the alloy can be obtained from

$$a_{C(gr)} = \frac{\left(p_{CO}^2 / p_{CO_2}\right)_{alloy}}{\left(p_{CO}^2 / p_{CO_2}\right)_{gr}} \tag{6.30}$$

In practice, a gas mixture with a known ratio of p_{CO}^2 / pCO_2 is passed over the ferrous alloy until equilibrium is attained and the concentration of C in the alloy then determined by chemical analysis.

6.2.1.3 Electrochemical Equilibria. An extension of the fundamental equation for Gibbs energy changes is necessary in order to allow for the fact that an electrochemical cell also does electrical work (in the surroundings). Just as with p–V work ($\delta w = -p\,dV$), electrical work can be expressed as an intensive quantity and its conjugate extensive property, namely, $\delta w = E\,dq$, where E is the electric potential and q the charge moved. The Gibbs energy form of the combined statement of the first and second laws for a system carrying out p–V, chemical, and electrical work is then

$$dG = S\,dT + V\,dp + \sum_i \mu_i\,dn_i + E\,dq \qquad (6.31)$$

An electrochemical cell is in equilibrium when no actual charge is flowing; when it is an open circuit cell, the electromotive force (emf) of which can be measured by using a very high impedance voltmeter.

At constant p and T and for an electrochemical cell in equilibrium, $dG|_{p,T} = 0$ and therefore

$$\sum_i \mu_i\,dn_i + E\,dq = 0 \qquad (6.32)$$

Since we wish to express E in volts and q in coulombs, a conversion factor is involved:

$$\sum_i \mu_i\,dn_i = -\sum_i z_i \mathcal{F}E\,dn_i \qquad (6.33)$$

where z_i is the charge on the ion being transported through the cell and $\mathcal{F}$ is the Faraday constant, being 96,494 when the μ_i are in joules and E, the emf of the cell, is in volts.

If only one species is being transported (at an infinitely slow rate), then for the difference in chemical potential between the two electrodes we can write:

$$\boxed{\Delta\mu_i = \mu_i' - \mu_i'' = -z_i \mathcal{F}E} \qquad (6.34)$$

In order to be able to use this equation, the electrochemical cell must have an electrolyte which transports only the ion of the alloy component under consideration and there must be no electron conduction in the electrolyte. Certain liquids, for example, some molten salts, and a few solids have this property.

As an example, an alkali metal chloride has been used as the electrolyte in the determination of the activity of Cd in liquid (Cd,Pb) alloys. Such a cell may be represented schematically as

$$Cd(l, T) \parallel CdCl_2 \text{ (in alkali chloride)} \parallel Cd\ (Cd_{1-x}Pb_x, l, T)$$

From (6.34) we see that the activity of Cd in the alloy can be determined from

$$a_{Cd}(l, T) = \exp\left[-\left(\frac{2\mathcal{F}E}{RT}\right)\right] \qquad (6.35)$$

since $E = 0$ when $x = 0$.

As was briefly mentioned in Chapter 2 for unary systems, another important source of experimental data, which has become important in the last decade, is the quantum mechanical calculation of the energies of formation of ordered compounds. The results from the best calculations are now of comparable accuracy to those which can be determined experimentally. Although the calculations refer to 0 K, it is possible, albeit with less precision at present, to also gain some information of the high-temperature properties by using appropriate models for the excitation processes. These first-principles calculations have the great advantage over experimental measurements in that it is just as easy to do calculations on unstable compounds as it is on the stable forms, whereas the experimentalist is usually constrained to carry out measurements on only the stable forms. Calculations on disordered alloy phases are more difficult at present, although by using supercells of certain ordered structures, approximations to the disordered state can be achieved.

EXERCISES

6.1 The following values for p_{Pb} over liquid (Pb,Sn) alloys at 1050 K have been reported:

Pb : $T_{fus} = 600.6$ K $\Delta_{fus} H° = 4799$ J mol^{-1}

x_{Pb}	0.879	0.737	0.657	0.514	0.397	0.282	0.176	0.091
$10^5 p$ /bar	4.253	4.101	3.889	3.621	2.929	2.311	1.555	1.015

Sn : $T_{fus} = 505.1$ K $\Delta_{fus} H° = 7029$ J mol^{-1}

(a) Calculate the activity, activity coefficient, partial molar Gibbs energy, and excess partial molar Gibbs energy of Pb relative to the Pb(l) reference state at 1050 K.

(b) Calculate the activity, activity coefficient, partial molar Gibbs energy, and excess partial molar Gibbs energy of Pb relative to a reference state of Pb(S) at 1050 K.

(c) Attempt to describe the data by a regular solution model and find the deviations of the data from this model.

(d) Plot $G^E_{Pb}/(1 - x_{Pb})^2$ versus x_{Pb} and carry out a regression analysis. Are the experimental data well described by a subregular solution model?

6.2 Assume that 1 mol of liquid nickel and 2 mol of liquid copper mix without an appreciable enthalpy and volume change. Is there a change in internal energy, U? In entropy, S? In Gibbs energy, G? Is any work done in this process if the mixing is carried out in a vacuum under an atmosphere of argon?

6.3 One mole of Cr(s) at 1600°C is added to a large quantity of (Fe,Cr) liquid solution (in which $x_{Fe} = 0.8$) which is also at 1600°C. If Fe and Cr form Raoultian ideal solutions, calculate the enthalpy and entropy changes in the solution resulting from the addition, assume the heat capacity difference between solid and liquid Cr is negligible.

Element	T_{fus}°/ K	$\Delta_{fus} H_m^{\circ}$/ kJ mol^{-1}
Cr	2130	16.93
Fe	1809	13.81

6.4 The activity coefficients of Ni [reference state Ni(l)] in liquid (Ni,Ti) alloys at 1700°C are:

x_{Ti}	1	0.9	0.8	0.7	0.6	0.5	0.4	0.3	0.2	0.1	0
γ_{Ni}	0.021	0.023	0.052	0.104	0.189	0.313	0.474	0.659	0.834	0.956	1

(a) Calculate a_{Ti} over the entire composition at intervals of 0.1 in x_{Ti}.
(b) Plot a_{Ti} and a_{Ni} as a function of composition.
(c) What are the Henry's law constants for Ni and Ti?

6.5 The excess Gibbs energies of bcc solid solutions of (Fe, Cr) and fcc solid solutions of (Fe, Cr) may be represented by the following expressions:

$$G^E(\text{bcc})/\text{J} = x_{Cr} x_{Fe} \, (25,104 - 11.7152T)$$

$$G^E(\text{fcc})/\text{J} = x_{Cr} x_{Fe} \, (13,108 - 31.823T + 2.748T \log_e T)$$

For the bcc phase:

(a) Calculate the partial Gibbs energy expressions for Fe and Cr.
(b) Plot the integral and partial Gibbs energies as a function of composition at 873 K.
(c) Plot a_{Cr} and a_{Fe} as a function of composition at 873 K.

6.6 The enthalpy of an alloy is often obtained by solution calorimetry since direct reaction of two elemental components may not always go to completion. One of the solvents used for metallic alloys is Sn since it has a low melting point but a high boiling point, making its vapor pressure low to quite high temperatures.

Two grams of Au, Cu, and $Au_{0.5}Cu_{0.5}$ at 298 K was individually dropped into a large bath of molten Sn maintained at 623 K. The chemical reaction equation is given by

$$M(s, 298 \text{ K}) = M(\text{in Sn(l)}, 623 \text{ K}, x_M \to 0)$$

where M refers to Au, Cu, or $Au_{0.5}Cu_{0.5}$. The enthalpies for the dissolution of M in the tin bath at 623 K were -142.1 J for Au, +614.6 J for Cu, and -176.3 J for $Au_{0.5}Cu_{0.5}$.

(a) Calculate the enthalpy of formation of $Au_{0.5}Cu_{0.5}$ at 298.15 K.

(b) Calculate the enthalpy of formation of $Au_{0.5}Cu_{0.5}$ at 623 K.

The values in the following table are required and refer to 1 mol of substance.

	T_{fus}/K	$\Delta_{fus} H^\circ /J$	$\langle C^\circ \rangle_p /J \text{ K}^{-1}$	A_r
Au	1336	12,550	26.0	197.2
Cu	1357	13,055	25.7	63.54
$Au_{0.5}Cu_{0.5}$	—	—	26.0	

6.7 (a) A stream of nitrogen gas is passed in a closed system over a boat containing mercury at 100°C. The flow rate of the nitrogen is slow enough to allow this gas to become saturated with mercury vapor. The total volume of nitrogen used is 22 liters measured at 20°C and 1 bar. The nitrogen was found to contain 0.0674 g of mercury. Calculate the vapor pressure of mercury at 100°C.

(b) When the same experiment is carried out with a sodium amalgam (Na+Hg) in which $x_{Na} = 0.122$, 22 liters of nitrogen gas saturated with mercury was found to contain 0.0471 g of mercury. Take pure mercury as the reference state and calculate the activity a, the activity coefficient γ_{Hg}, and the difference $\mu_{Hg} - \mu_{Hg}^\circ$ in this amalgam.

6.8 (a) Will a gas mixture containing 97% H_2O and 3% H_2 oxidize nickel at 1000 K?

(b) An alloy containing 10 at % Ni and 90 at % Au is a solid solution to 1000 K. It is found that this solution reacts with water vapor to form NiO. Assume that approximate measurements indicate that the reaction reaches equilibrium when the water vapor–hydrogen mixture contains 0.35% hydrogen by volume. Find the corresponding value of the activity coefficient γ of Ni in the alloy.

6.9 A pure iron wire is exposed to an atmosphere composed of 95% CO and 5% CO_2 at a total pressure of 1 bar and 1200 K. When equilibrium is established, the carbon content of the wire is found to be 0.4 wt %, $A_r(C)$ = 12.01, and $A_r(Fe) = 55.84$.

(a) Calculate a_C with reference to graphite as the reference state.

(b) If the experiment is reported at the same temperature and same gas composition but with the total gas pressure of 2 bars, estimate the equilibrium carbon concentration in the wire. State any assumptions made.

(c) Solutions of carbon in solid iron exhibit positive deviations from Henry's law. Given the information, is your answer to part (b) too high or too low?

7 Thermodynamics of Binary Solutions III: Experimental Results and Their Analytical Representation

7.1 SOME EXPERIMENTAL RESULTS

For an alloy to be stable, at a given p and T, with respect to the component elements in the same structural form at a particular composition, $\Delta_{\mathrm{mix}} G_m < 0$. If this were not the case, a mechanical mixture of the perfectly pure metal components would be stable at that composition.

It is possible, however, to obtain negative $\Delta_{\mathrm{mix}} G_m$ values from different combinations of the contributory $\Delta_{\mathrm{mix}} H_m$ and $\Delta_{\mathrm{mix}} S_m$ values:

(i) If $\Delta_{\mathrm{mix}} H_m > 0$, then $\Delta_{\mathrm{mix}} G_m < 0$ only if $\Delta_{\mathrm{mix}} S_m > \Delta_{\mathrm{mix}} H_m / T$.

(ii) If $\Delta_{\mathrm{mix}} H_m \ll 0$, then $\Delta_{\mathrm{mix}} G_m$ can be < 0 not only for $\Delta_{\mathrm{mix}} S_m > 0$ but perhaps also when $\Delta_{\mathrm{mix}} S_m < 0$.

Experimental results for the thermodynamic properties of real binary alloys show these different ways in which $\Delta_{\mathrm{mix}} G_m$ can be < 0.

7.1.1 Liquid Alloys

A cursory glance at any compilation of phase diagrams reveals that liquid metals are much more likely to form a complete series of liquid alloy solutions than is the case for solid alloys. Immiscibility and compound formation are much more likely to be found in solid alloys.

Some results for $\Delta_{\mathrm{mix}} H_m$ and $\Delta_{\mathrm{mix}} S_m$ for liquid alloys are presented typical in Figures 7.1 and 7.2.

Figure 7.1 shows an example for liquid (Al,Ti) alloys, where $\Delta_{\mathrm{mix}} H_m < 0$, and one for liquid (Pb,Zn) alloys, where $\Delta_{\mathrm{mix}} H_m > 0$. Note that the magnitude

Materials Thermodynamics. By Y. Austin Chang and W. Alan Oates
Copyright 2010 John Wiley & Sons, Inc.

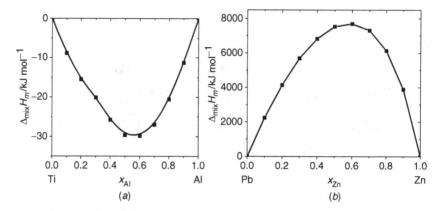

Figure 7.1 $\Delta_{\mathrm{mix}} H_m > 0$ for two liquid alloys: (a) (Ti,Al) alloys at 2000 K; (b) (Pb,Zn) alloys at 926 K.

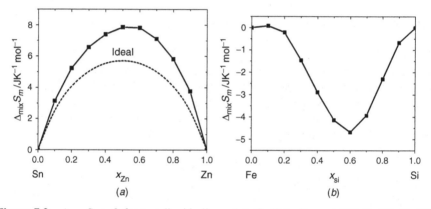

Figure 7.2 $\Delta_{mix} S_m > 0$ for two liquid alloys: (a) (Sn,Zn) alloys at 800 K; (b) (Fe,Si) alloys at 1853 K.

of $\Delta_{\mathrm{mix}} H_m$ for (Al,Ti) is considered to be fairly large for alloy mixing, although the magnitudes are much lower than can be found for ionic compound formation, for example.

Figure 7.2 shows an example for liquid (Sn,Zn) alloys, where $\Delta_{\mathrm{mix}} S_m > 0$, and one for liquid (Fe,Si) alloys, where $\Delta_{\mathrm{mix}} S_m < 0$. The values of $\Delta_{\mathrm{mix}} S_m^{\mathrm{id}}$ for an ideal solution are also shown in Figure 7.2a. It can be seen that the results shown for (Fe,Si) represent a large negative deviation from that for an ideal solution, that is, $S_m^{\mathrm{E}} (= \Delta_{\mathrm{mix}} S_m - \Delta_{\mathrm{mix}} S_m^{\mathrm{id}})$ is large and negative.

The results for S_m^{E} from a compilation for 107 liquid binary alloys are shown as a histogram in Figure 7.3. It is apparent that S_m^{E} often departs significantly from zero, that is, many binary liquid alloys have entropies of mixing which deviate significantly from that of an ideal solution. It can also be seen that S_m^{E} is more often negative than positive and that, in many cases, S_m^{E} is sufficiently large

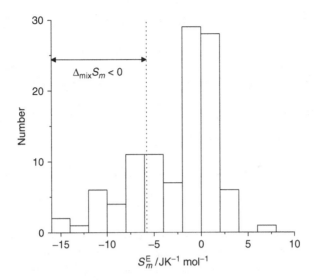

Figure 7.3   Histogram of S_m^E for some liquid binary alloys. (Results from *Journal of Alloys and Compounds*, **325** (2001) 118).

and negative that $\Delta_{mix}S_m$ is itself negative (the dashed vertical line corresponds with $S_m^E = -\Delta_{mix}S_m$ at the midcomposition for an ideal solution).

7.1.2   Solid Alloys

Alloy solution phases which exist over the whole composition range are not very common in the case of solid alloys. These continuous series of solid solutions occur when the magnitudes of the values for $\Delta_{mix}H_m$ are relatively small. Two examples, one for (Ag,Pd) where $\Delta_{mix}H_m < 0$ and another for (Au,Ni) where $\Delta_{mix}H_m > 0$, are shown in Figure 7.4a. In those solid alloys which do form a continuous series of solid solutions it is also found that, because $\Delta_{mix}H_m$ is necessarily small, S_m^E is close to zero. On this point, as shown in Figure 7.5, there appears to be a correlation between $\Delta_{mix}H_m$ and S_m^E for liquid and solid solutions.

A more common occurrence with solid alloys is for them either to be immiscible or to form intermediate phases.

Phase separation will occur below some temperature whenever $\Delta_{mix}H_m > 0$ because of the decreasing influence of temperature on the $T\Delta_{mix}S_m$ term, that is, $\Delta_{mix}G_m$ will eventually become > 0 if $\Delta_{mix}H_m > 0$. In the case of the (Au,Ni) system shown in Figure 7.4b, phase separation occurs at temperatures below about 1083 K.

Intermediate-phase formation occurs whenever $\Delta_{mix}H_m$ is large and negative, indicative of a strong chemical affinity between the two metals involved. In such cases, several different intermediate phases are often found to occur and these will most likely be based on different parent phase structures; that is, some will be based on an fcc parent structure, others on a bcc parent. This means that it

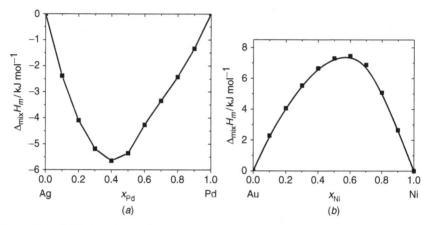

Figure 7.4 Differing signs of $\Delta_{\text{mix}} H_m$ for two solid solution alloys: (*a*) fcc (Ag,Pd) alloys at 1000 K, have small exothermic $\Delta_{\text{mix}} H_m$; (*b*) fcc (Au,Ni) alloys at 1153 K, have small endothermic $\Delta_{\text{mix}} H_m$.

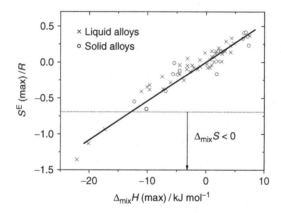

Figure 7.5 Kubaschewski correlation between excess entropies and enthalpies of mixing for some liquid and solid solution phases.

is no longer possible to plot $\Delta_{\text{mix}} H_m$ or $\Delta_{\text{mix}} S_m$ as a function of composition as we did for the case when a continuous series of solid solutions are formed. Nevertheless, it is possible to observe the general trend by plotting $\Delta_f H_m$ or $\Delta_f S_m$ with respect to the stable element reference states at that temperature. An example is given in Figure 7.6 for solid (Al,Ni) alloys. The reference states here are Al(l) and Ni(fcc). The solid lines where there are no points represent the two phase regions (at 1000 K). Very large negative $\Delta_f S_m$ values are indicated in this system and this conclusion would not change markedly were reference states selected for each intermediate phase which are appropriate for calculating $\Delta_{\text{mix}} S_m$ for that particular phase.

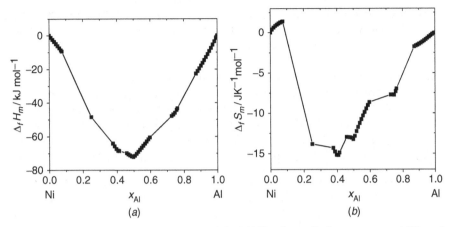

Figure 7.6 Properties at 1000 K for solid (Al,Ni) alloys. Reference states Al(l) and Ni(fcc): (*a*) $\Delta_f H_m$; (*b*) $\Delta_f S_m$.

Just as is apparent from the histogram shown in Figure 7.3 for liquid alloys, a similar plot for solid alloys also indicates that negative $\Delta_f S_m$ values are often found in solid alloys.

Summary   The experimental results for the thermodynamic properties of real binary alloys indicate that:

(i) Both positive (endothermic) and negative (exothermic) values for $\Delta_{mix} H_m$ or $\Delta_f H_m$ are found in both liquid and solid alloys.

(ii) Deviations of the entropies of mixing from $\Delta_{mix} S_m^{id}$ are commonplace.

(iii) Both positive and negative values for S_m^E are found, although negative S_m^E values are more common. Frequently, negative deviations from ideal mixing can be so large that $\Delta_{mix} S_m$ and $\Delta_f S_m$ are negative.

(iv) Asymmetry of the properties as a function of composition is frequently found.

These observations concerning the thermodynamic properties of liquid and solid binary alloys must be taken into account when attempting to atomistically model these properties.

7.2 ANALYTICAL REPRESENTATION OF RESULTS FOR LIQUID OR SOLID SOLUTIONS

Principal reasons for having analytical representations are:

1. It provides the means for obtaining efficient storage solution-phase properties in computer databases.

2. The equations for the integral and partial molar properties are thermodynamically self-consistent. Thus it is unnecessary, for example, to carry out any Gibbs–Duhem integrations.

The principal reason for defining and having analytical representations of the excess thermodynamic functions comes from the fact that, in calculating them, the logarithmic terms are removed from the integral and partial molar expressions for $\Delta_{\text{mix}}S_m$ and $\Delta_{\text{mix}}G_m$. With these removed, there is the possibility of obtaining simple polynomial expressions for the excess functions.

Margules first suggested a polynomial representation of $H_m^E (= \Delta_{\text{mix}}H_m$ since $\Delta_{\text{mix}}H_m^{\text{id}} = 0)$:

$$H_m^E = x_A x_B \left[\lambda_0 + \lambda_1 x_B + \lambda_2 x_B^2 + \lambda_3 x_B^3 + \cdots \right]$$

$$= x_A x_B \sum_{i=0}^{n} \lambda_i x_B^i \tag{7.1}$$

It can be seen that this form of representation has the required limiting value of zero at both end compositions. A polynomial of any required order in x_B can be used in the summation term and the λ's may or may not be temperature dependent.

More recently, (7.2) has been assumed to apply to G_m^E rather than to just H_m^E:

$$G_m^E = x_A x_B \sum_{i=0}^{n} \lambda_i x_B^i \tag{7.2}$$

Polynomials for G_m^E which differ slightly from the one given in (7.2) are possible. One which is frequently used in computer databases, because it is more readily extended for use with multicomponent solutions than (7.1), is known as the *Redlich–Kister* equation:

$$G_m^E = x_A x_B \sum_{j=0}^{n} L_j (x_A - x_B)^j \tag{7.3}$$

In this equation the L_j parameters may or may not have to be assumed to be temperature dependent in describing a set of experimental results.

The expressions for the partial molar quantities corresponding with (7.2) are

$$\mu_A^E = x_B^2 \sum_{j=0}^{n} L_j (x_A - x_B)^{j-1} [(2j + 2)x_A - 1] \tag{7.4}$$

$$\mu_B^E = x_A^2 \sum_{j=0}^{n} L_j (x_A - x_B)^{j-1} [1 - x_B(2j + 2)] \tag{7.5}$$

A Raoultian ideal solution is one where all the L_j parameters are zero. When only the term with $n = 0$ is required to describe the experimental results, then the integral and partial molar quantities are given by

$$G_m^E = L_0 x_A x_B \tag{7.6}$$

$$\Delta_{\text{mix}} G_m = L_0 x_A x_B + RT(x_A \log_e x_A + x_B \log_e x_B) \tag{7.7}$$

$$\mu_A^E = L_0 x_B^2 \tag{7.8}$$

$$\Delta_{\text{mix}} \mu_A = L_0 x_B^2 + RT \log_e x_A \tag{7.9}$$

$$\mu_B^E = L_0 x_A^2 \tag{7.10}$$

$$\Delta_{\text{mix}} \mu_B = L_0 x_A^2 + RT \log_e x_B \tag{7.11}$$

A solution whose thermodynamic properties can be adequately described by these equations is called a *regular solution*. If the experimental results can be described without L_0 being temperature dependent, then the solution is said to be *strictly regular*. In this case, the first terms in (7.7), (7.9), and (7.11) are enthalpies and second terms are entropies. Because the strictly regular solution is simple to handle and because it is occasionally not all that far away from providing a reasonable representation of the properties of some real alloys, it is often used in illustrative phase diagram calculations. Note that while a regular solution has a constant value of $G_m^E/x_A x_B$, the converse is not true. It is possible for $G_m^E/x_A x_B$ to be a composition-independent constant even when $\Delta_{\text{mix}} S_m \neq \Delta_{\text{mix}} S_m^{\text{id}}$.

When the properties of a solution are not well represented by using just the L_0 terms, additional terms are added in (7.3). Those solutions which can be represented by using only the L_0 and L_1 terms have often been called *subregular* and those requiring $L_0, L_1,$ and L_2 sub-subregular. The sketch in Figure 7.7

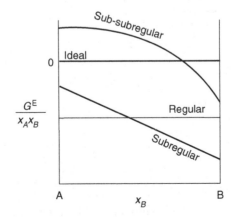

Figure 7.7 Behavior of $G_m^E/x_A x_B$ for different numbers of coefficients used in Redlich–Kister equation. No coefficients—ideal solution; L_0 only—regular solution; L_0, L_1 only—subregular solution; L_0, L_1, L_3 only—sub-subregular solution.

shows how $G_m^E / x_A x_B$ varies as a function of composition with an increasing number of L_j parameters together with the names usually given.

The use of high-order polynomials is not recommended in the fitting of experimental results. It would be rare to go beyond the use of L_3, since the use of further terms increases the likelihood of finding waves in the polynomial representation of sparse data. If a solution's properties cannot be fitted by a low-order polynomial, then it is time to look for another type of representation which is based more on the physicochemical nature of the solution. This is the situation, to be discussed in later chapters, when wishing to describe the thermodynamic properties of intermediate phases.

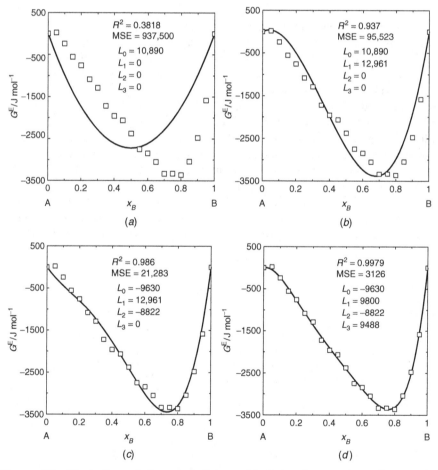

Figure 7.8 Illustration of improved fitting with increasing number of terms in Redlich–Kister equation: (a) L_0 parameter only used in fitting; (b) L_0 and L_1 parameters used in fitting; (c) $L_0, L_1,$ and L_2 parameters used in fitting; (d) $L_0, L_1, L_2,$ and L_3 parameters used in fitting.

Example 7.1 Using Redlich–Kister Equation to Describe Experimental Results

A satisfactory analytical description of the excess thermodynamic properties requires the least-squares fitting of the Redlich–Kister equation to the experimental results. A decision has to be made as to how many parameters in this equation should be used in the description. This is a standard statistical problem and, without going into any detail, we may illustrate what is involved by considering the experimental results shown in Figure 7.8. The results available in this example are more plentiful than is normally the case for real experimental results: this having been done purposely in order to illustrate the procedure.

One can start off by using only the L_0 term in (7.3). It is immediately obvious in Figure 7.8a that the description using this single parameter is very poor. This is also apparent from the low value of the coefficient of determination, R^2 statistic (gives the probability of variability in a data set accounted for by the model), and the high value of the mean-square error (MSE) (a measure of accuracy which is computed by squaring the individual errors for each item in the data, then finding the average of the sum of those squares).

The remaining figures in Figure 7.8 show the effect of increasing the number of L parameters in the description. In this case, the values of R^2 and the MSE have reached acceptable values only when the $L_0, L_1, L_2,$ and L_3 parameters are used, a fact which is also obvious by visual inspection.

Experimental results for the thermodynamic properties of a solution phase are rarely as plentiful as the ones shown in Figure 7.8. When the data are sparse, it is dangerous to use a third-order Redlich–Kister polynomial to describe the results. This is illustrated in Figure 7.9. The fitted curve shown has included the $L_0, L_1, L_2,$ and L_3 parameters and it can be seen that waves have already started to appear in the least-squares curve. When it is recalled that it is the derivative of this fitted curve which is required in the calculation of the partial molar quantities, the dangers are obvious.

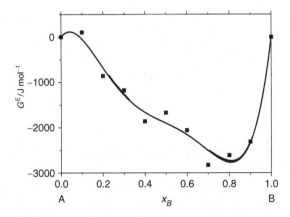

Figure 7.9 When the data are sparse, it is dangerous to use a high-order Redlich–Kister polynomial to describe the solution properties.

EXERCISES

7.1 When $\mu_B^E = L_0(1 - x_B)^2$, show by the Gibbs–Duhem relationship that $\mu_A^E = L_0(1 - x_A)^2$.

7.2 Liquid solutions of (Mg,Al) exist from pure Mg(l) to Al(l) at 700°C. Estimate the partial pressure of Mg over an alloy containing 90 at % Al, 700°C, making any assumption necessary. Use the following equations for the vapor pressures of Al(l) and Mg(l):

$$\text{Al(1933–2793 K):} \quad \log_{10}\left(\frac{p}{\text{Pa}}\right) = -\frac{16,380}{T} - 1.0 \log_{10} T + 14.445$$

$$\text{Mg(1922–1363 K):} \quad \log_{10}\left(\frac{p}{\text{Pa}}\right) = -\frac{7550}{T} - 1.41 \log_{10} T + 14.915$$

7.3 The results for liquid (Cu,Sn) alloys shown in Figure 5.3 are described by the following equation:

$$\Delta_{\text{mix}} H_m = x_{\text{Cu}} x_{\text{Sn}}[-90028.0 - 5, 8381T$$
$$- (20100.4 - 3.6366T)(x_{\text{Cu}} - x_{\text{Sn}})$$
$$- 10528.4(x_{\text{Cu}} - x_{\text{Sn}})^2] \quad \text{J/mol}$$

Derive the equation for $\Delta_{\text{mix}} H_{\text{Sn}}$ and compare the results with those shown in Figure 5.4.

7.4 Show that the coefficients used in (7.1) and (7.3) are related to one another. The enthalpy of mixing of (Cu,Ni) alloys exhibiting the fcc structure can be described approximately by the equation

$$\Delta_{\text{mix}} H_m = 8400 x_{\text{Ni}}(1 - x_{\text{Ni}}) \quad \text{J/mol} \quad (7.12)$$

(a) Calculate the integral enthalpy of mixing with 30 mol % of Ni.
(b) Calculate the partial enthalpy of mixing of Cu and Ni in an alloy with 30 mol % of Ni.

7.5 The partial molar enthalpy of mixing of Ge in (Si,Ge) alloys can be represented by the equation

$$\Delta_{\text{mix}} H_{\text{Ge}} = 5820(1 - x_{\text{Ge}})^2 \quad \text{J/mol} \quad (7.13)$$

(a) Calculate the partial enthalpy of mixing of Ge in a (Si,Ge) alloy with 40 mol % Ge.
(b) Calculate the partial enthalpy of mixing of Si in the same alloy.
(c) Calculate the integral enthalpy of mixing of (Si,Ge) for the same alloy.

8 Two-Phase Equilibrium I: Theory

8.1 INTRODUCTION

We have seen that the most stable equilibrium state in a closed system at constant p and T coincides with where the Gibbs energy is at its global minimum value. But we have also seen that it is possible to have local equilibrium points which exist in a metastable state. Liquid water existing below $0°C$ is an example of a metastable phase in a one-component system.

Metastable states are often encountered in alloys and it is important to appreciate that these local equilibria are just as amenable to the application of thermodynamics as is the stable state; that is, thermodynamics is applicable to situations where any change of the system with time is imperceptible in the time window of our observations. For example, in a plain carbon steel, we may have metastable martensite at room temperature; its decomposition rate is imperceptible. The usual thermodynamic relations can be applied and thermodynamic properties assigned to martensite in this state. On tempering the steel at a higher temperature, it starts to decompose (the system is changing with time) and thermodynamics is not applicable to this state. When the martensite is completely decomposed to ferrite and carbide and the system is once more effectively time independent, thermodynamics is again applicable. However, another transition state can then be obtained if the alloy is annealed at a higher temperature to give the final global equilibrium of ferrite plus graphite. Thermodynamic relations are not applicable during the transition stage, except in providing hints for the direction of change. Thermodynamics is again applicable, in a quantitative way, to the final global equilibrium.

In the following, we are concerned with finding solutions to a more restricted problem, namely, locating the equilibrium position, metastable or stable, between two *prespecified* phases. The equilibrium conditions for this situation can be derived in a straightforward way. Just as we did for unary systems, we seek a constrained minimization of the Gibbs energy by using Lagrangian multipliers.

Materials Thermodynamics. By Y. Austin Chang and W. Alan Oates
Copyright 2010 John Wiley & Sons, Inc.

8.2   CRITERION FOR PHASE EQUILIBRIUM BETWEEN TWO SPECIFIED PHASES

8.2.1   Equilibrium between Two Solution Phases

Consider two binary solution phases, α and β, in contact with one another as shown in Figure 8.1. The total system comprising the two phases is closed, but each subsystem, namely, each individual phase, is open to the other.

The Gibbs energy in each phase is given by the weighted sum of the chemical potentials, that is, $G = \sum_i n_i \mu_i$, and the total Gibbs energy of the two-phase system by $G = \sum_j \sum_i n_i^j \mu_i^j$.

We can write the constraints on the Gibbs energy minimization and give an associated Lagrangian multiplier as follows:

$$\lambda_1 \quad \text{for} \quad n_A(\text{total}) - n_A^\alpha - n_A^\beta = 0 \tag{8.1}$$

$$\lambda_2 \quad \text{for} \quad n_B(\text{total}) - n_B^\alpha - n_B^\beta = 0 \tag{8.2}$$

and the Lagrangian function $\mathcal{L}$ to be minimized becomes

$$\mathcal{L} = G + \lambda_1(n_A - n_A^\alpha - n_A^\beta) + \lambda_2(n_B - n_B^\alpha - n_B^\beta) \tag{8.3}$$

$$= n_A^\alpha \mu_A^\alpha + n_A^\beta \mu_A^\beta + n_B^\alpha \mu_B^\alpha + n_B^\beta \mu_B^\beta \tag{8.4}$$

$$+ \lambda_1(n_A - n_A^\alpha - n_A^\beta) + \lambda_2(n_B - n_B^\alpha - n_B^\beta) \tag{8.5}$$

Partial differentiation with respect to, for example, n_A^α gives the minimum in $\mathcal{L}$:

$$\frac{\partial \mathcal{L}}{\partial n_A^\alpha} = \mu_A^\alpha - \lambda_1 = 0 \tag{8.6}$$

$$\frac{\partial \mathcal{L}}{\partial n_A^\beta} = \mu_A^\beta - \lambda_1 = 0 \tag{8.7}$$

$$\frac{\partial \mathcal{L}}{\partial n_B^\alpha} = \mu_B^\alpha - \lambda_2 = 0 \tag{8.8}$$

$$\frac{\partial \mathcal{L}}{\partial n_B^\beta} = \mu_B^\beta - \lambda_2 = 0 \tag{8.9}$$

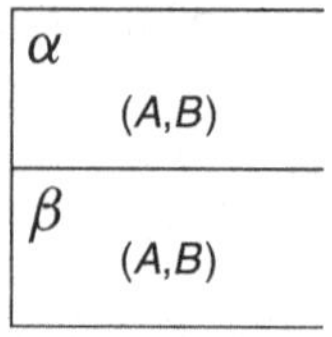

Figure 8.1   The two solution phases α and β are in open contact with one another but the $\alpha + \beta$ system is closed.

Elimination of the Lagrangian multipliers leads to the condition for the constrained minimum in G, that is, the condition of equilibrium:

$$\mu_A^\alpha = \mu_A^\beta \tag{8.10}$$

$$\mu_B^\alpha = \mu_B^\beta \tag{8.11}$$

In general, the condition of equilibrium between any two prespecified phases α and β is

$$\boxed{\mu_i^\alpha = \mu_i^\beta} \quad \text{for all components } i \text{ in phases } \alpha \text{ and } \beta \tag{8.12}$$

The physical meaning of the Lagrangian multipliers used for the mass balance constraints λ_1 and λ_2 is apparent from (8.6)–(8.9). They are seen to be chemical potentials.

Since $\mu_i = f\left(p, T, x_i\right)$, then solution of the set of equations (8.12) yields the compositions of the phases in equilibrium at a given p and T. The meaning of (8.12) is shown graphically in Figure 8.2. At the constant p and T chosen, Figure 8.2a shows the Gibbs energies for the two phases. Starting from the A-rich side, it can be seen that the α-phase has the lower Gibbs energy until the composition x_B^α is reached. From there until the composition x_B^β, the two-phase mixture of $\alpha + \beta$ has a lower Gibbs energy than either of the single phases. After x_B^β the β-phase becomes the stable state. Figure 8.2b summarizes this information in a tie-line of the $T - x_i$ phase diagram. There is another important result from

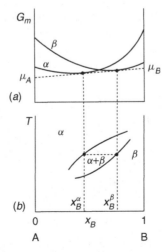

Figure 8.2   Equilibrium between two phases in Figure 8.1.

the calculation. The constraint $n_A = n_A^\alpha + n_A^\beta$ can be rewritten as

$$\frac{n_A}{N} = \frac{n_A^\alpha}{N} + \frac{n_A^\beta}{N} = \frac{N_\alpha}{N} \cdot \frac{n_A^\alpha}{N_\alpha} + \frac{N_\beta}{N} \cdot \frac{n_A^\beta}{N_\beta} \tag{8.13}$$

which is equivalent to

$$\boxed{x_A = f^\alpha x_A^\alpha + f^\beta x_A^\beta} \tag{8.14}$$

which is a statement of the Lever rule. Not only are the compositions of the phases in equilibrium obtained but also their relative amounts.

We see, then, that the constrained Gibbs energy minimization for two prespecified phases in equilibrium in a closed system of given overall composition at constant p and T gives us two important results:

(i) The phase *compositions* from $\mu_i^i = \mu_i^j$
(ii) The phase *fractions* from $x_i = f^i x_i^i + f^j x_i^j$

It is emphasized again that (8.12) is a necessary but not a sufficient condition for calculating the stable or global equilibrium in a system. This is illustrated graphically in Figure 8.3. There, the tangent intercepts to both pairs of points *ab* and *cd* fulfill the condition given in (8.12), but it is clear that, for an alloy with composition between points *b* and *c*, only one of them, the tangent to the points *ab* refers to the global equilibrium. Remember also that the equilibrium between two bulk phases at constant p and T is being discussed. We have not considered any effects due to stresses, external magnetic fields, and so on.

In this case of simple binary systems we can look at the phases singly and two at a time, using (8.12) to obtain the compositions of phases in equilibrium. It is then possible to determine the location of the global equilibrium by finding which phase or combination of phases has the lowest G. The calculation of

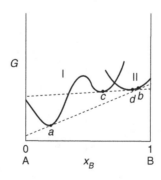

Figure 8.3 Illustration of difference between global and local minimum in G for two-phase mixtures.

multicomponent, multiphase phase diagrams requires a more sophisticated algorithm.

8.2.2 Equilibrium between a Solution Phase and a Stoichiometric Compound Phase

To some extent this calculation is simpler than the one discussed above in that there is only one unknown, the solution-phase composition, since the composition of the stoichiometric phase is given. The Gibbs energies involved are sketched in Figure 8.4. The Gibbs energy of the stoichiometric compound is a point on this $G - x_B$ diagram. Although a tangent cannot be drawn to a point, it is clear from the figure that there is a unique solution to the problem.

Suppose that the β-phase is a stoichiometric compound, $A_m B_n$. The need to maintain stoichiometry introduces another constraint into the Gibbs energy minimization over those used for the two-solution-phase case. The extra Lagrangian multiplier is

$$\lambda_3 \quad \text{for} \quad mn_B^\beta - nn_A^\beta = 0 \tag{8.15}$$

Minimization of the Lagrangian function using (8.1), (8.2), and (8.15) gives

$$\frac{\partial \mathcal{L}}{\partial n_A^\alpha} = \mu_A^\alpha - \lambda_1 = 0 \tag{8.16}$$

$$\frac{\partial \mathcal{L}}{\partial n_A^\beta} = \mu_A^\beta - \lambda_1 - n\lambda_3 = 0 \tag{8.17}$$

$$\frac{\partial \mathcal{L}}{\partial n_B^\alpha} = \mu_B^\alpha - \lambda_2 = 0 \tag{8.18}$$

$$\frac{\partial \mathcal{L}}{\partial n_B^\beta} = \mu_B^\beta - \lambda_2 + m\lambda_3 = 0 \tag{8.19}$$

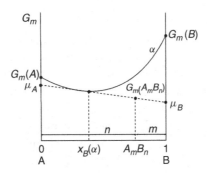

Figure 8.4 Equilibrium between solution phase, α, and stoichiometric compound phase, $A_m B_n$.

Elimination of the Lagrangain multipliers from these equations yields the condition of equilibrium:

$$\boxed{m\mu_A^\beta + n\mu_B^\beta = G_m(A_m B_n) = m\mu_A^\alpha + n\mu_B^\alpha} \tag{8.20}$$

The graphical interpretation of this equation is clear from Figure 8.4. It can be seen that $G_m(A_m B_n)$ is given by the composition-weighted sum of the chemical potentials in the α-phase when the same reference states are used for both solution and compound phases.

In discussing stoichiometric phases, it should be appreciated that all intermediate phases are nonstoichiometric to some extent above 0 K since, as well as vibrational excitations, atomic disorder excitations (defects) are always present. With a compound like GaAs, for example, the deviation from stoichiometry is less than one part in 10,000 even close to its melting point. But its important technological properties, that is, its electrical properties, are extremely sensitive to these very small deviations. With such small deviations from stoichiometry, we are usually justified in using the approach discussed above.

In the case of some alloy intermediate phases, however, the deviation from stoichiometry can be quite large (on the order of $\Delta x = 0.1$). For such phases, the above treatment is inadequate. But it is also found that a low-order polynomial representation for G^E, as we have used for solution phases, is also unsatisfactory for nonstoichiometric intermediate phases. It is necessary to look at the physical modeling of these intermediate phases in order to obtain a more reliable representation for their Gibbs energy. This is left for later chapters.

8.3 GIBBS'S PHASE RULE

The phase rule is concerned with evaluating the number of independent *field variables* required to specify a state. In the following we evaluate the number of degrees of freedom (independent variables) for $p - V$ work systems. If we have more than just $p - V$ work, then the number of independent variables increases and the following derivation is no longer correct. The following derivation is very similar to that already given for unary systems.

(A) *Total Number of Field Variables*   For C components the field variables are $p, T, \mu_1, \mu_2, \ldots, \mu_C$. The total number of variables for each phase is $C + 2$ and the total number of variables for ϕ-phases is $\phi(C + 2)$.

(B) *Number of Constraints on Field Variables*   The field variables are not all independent:

(a) $T^{(1)} = T^{(2)} = \cdots = T^{(\phi)}$ $\phi - 1$ relations

(b) $p^{(1)} = p^{(2)} = \cdots = p^{(\phi)}$ $\phi - 1$ relations

(c) $\mu_i^{(j)}$ constraints R_μ relations

Here, R_μ is the number of chemical potential constraints. The total number of constraints equals $2(\phi - 1) + R_\mu$.

(C) *Number of Independent Variables, F*   This is given by the total number of variables minus the number of constraints, $\phi(C + 2) - 2(\phi - 1) - R_\mu$, which gives

$$\boxed{F = \phi C + 2 - R_\mu} \quad p - V \text{ work only} \tag{8.21}$$

which is the general form of Gibbs's phase rule for systems undergoing $p - V$ work only. At constant total pressure we lose one degree of freedom and the phase rule for $p - V$ work only becomes

$$F = \phi C + 1 - R_\mu \quad \text{fixed pressure} \tag{8.22}$$

Let us now consider the number of chemical potential constraints, R_μ. If we assume that all species are mobile between all the ϕ-phases, then

$$\mu_i^{(1)} = \mu_i^{(2)} = \cdots = \mu_i^{\phi} \tag{8.23}$$

Thus there are $C(\phi - 1)$ constraints on the chemical potentials for C components. But we must also remember that the constant p, T Gibbs–Duhem equation, $\sum_i n_i \, d\mu_i = 0$, places an extra constraint on the various μ_i in each phase; that is, there are ϕ more constraints to include. Thus

$$R_\mu = C(\phi - 1) + \phi \tag{8.24}$$

Substituting (8.24) into (8.21), the phase rule becomes

phase equilibrium,

$$\boxed{F = C - \phi + 2} \quad p - V \text{ work only,} \tag{8.25}$$

all species mobile in all phases

If the total pressure is held constant, then one independent variable is lost and we obtain the most frequently encountered form of the phase rule for use in phase diagram work:

phase equilibrium, fixed pressure,

$$\boxed{F = C - \phi + 1} \quad p - V \text{ work only,} \tag{8.26}$$

all species mobile in all phases

Using (8.26), we see that, for a single phase ($\phi = 1$) in a binary system ($C = 2$), $F = 2$. This means that, as we have already seen, all the information for this system can be plotted on a two-dimensional phase diagram. When two phases are present, $F = 1$, which means that, if the temperature is specified (at constant total pressure), everything else is fixed, that is, the compositions of the coexisting phases.

It is important to appreciate the assumptions used in arriving at this most commonly used form of the phase rule, (8.26). As well as assuming that we are dealing with $p - V$ work only systems, we assume that all the species are *mobile* between all phases and that no chemical reactions are occurring within the system. When any such constraint comes into play, (8.26) is no longer applicable.

Note that the number of degrees of freedom specifies the number of independent variables which are required to fix all the *field* properties of the system. If we wish to fix the *extensive* properties, then it is necessary to also fix an extensive property so that the size of each phase in the system is fixed.

EXERCISES

8.1 Use the Lagrangian function approach to derive the equilibrium condition between a solution phase, (A,B), and the pure metal B.

8.2 Given the experimental data below for G_m^E/J mol^{-1} for some binary liquid alloys, decide, for each alloy, the number of terms which should be used in a Redlich–Kister representation.

x_B	Ag–Au 1500 K	Fe–Ti 2000 K	Cu–Zn 1500 K	Al–Ca 1000 K
0	0	0	0	0
0.05	−697.87	−2480.87	−1150.61	−2517.21
0.1	−1322.28	−4658.02	−2176.18	−4987.05
0.15	−1873.23	−6538.54	−3069.88	−7332.12
0.2	−2350.72	−8129.54	−3826.62	−9486.69
0.25	−2754.75	−9438.09	−4443.11	−11396.1
0.3	−3085.32	−10471.3	−4917.78	−13016.2
0.35	−3342.43	−11236.3	−5250.87	−14312.6
0.4	−3526.08	−11740.1	−5444.33	−15260
0.45	−3636.27	−11989.9	−5501.91	−15841.8
0.5	−3673	−11992.8	−5429.11	−16049.2
0.55	−3636.27	−11755.7	−5233.19	−15880.8
0.6	−3526.08	−11286	−4923.18	−15341.8
0.65	−3342.43	−10590.5	−4509.85	−14443.2
0.7	−3085.32	−9676.51	−4005.77	−13201.8
0.75	−2754.75	−8551.03	−3425.23	−11638.8
0.8	−2350.72	−7221.18	−2784.32	−9779.66
0.85	−1873.23	−5694.06	−2100.86	−7653.37
0.9	−1322.28	−3976.76	−1394.45	−5291.72
0.95	−697.87	−2076.37	−686.46	−2728.73
1	0	0	0	0

8.3 The following sparse experimental values for G^E for a liquid alloy (A,B) have been obtained. Fit them to a Redlich–Kister equation using $L_0, \ldots, L_4$ parameters and plot your results. Now use the fitted equation to obtain μ_B as a function of x_B. What do you conclude?

x_B	0	0.1	0.2	0.3	0.4	0.5	0.6	0.7	0.8	0.9	1
G^E/kJ mol^{-1}	0	−4.55	−8.95	−11.82	−12.63	−14.4	−15.34	−12.19	−7.76	−2.79	0

8.4 Equation (8.26), for the constant-total-pressure form of the phase rule, was derived on the basis of assuming that all species are mobile between all phases. Consider the case of the two-phase equilibrium $(A, B, C)/(A, B)$, that is, where species C is not mobile between the two phases and examine whether any modifications to the previous phase rule are necessary.

9 Two-Phase Equilibrium II: Example Calculations

In this chapter we will be concerned with some specific examples which have been chosen to illustrate the calculation of the equilibrium between two prespecified phases in a binary system. Such calculations may be carried out by following these steps:

(A) *Introduce Thermodynamic Relations* The condition of equilibrium between two *prespecified* phases, α and β, assuming that all species are mobile between both phases, is given by

$$\mu_i^\alpha = \mu_i^\beta \qquad \forall i \qquad (9.1)$$

We will always use the MS&E approximation where the thermodynamic properties of condensed phases are assumed to be independent of the total pressure.

(B) *Select Reference States* In order to be able to use (9.1) we must choose the same reference state for a particular *component* in both phases. Computer databases use the SER state for all substances in all phases, so that this requirement is fulfilled. Consider, for example, pure Fe(fcc) at temperature T. Its standard molar enthalpy at a temperature T, using SER, is given as

$$H_m^\circ(\text{Fe(fcc)},\text{SER}, T)$$
$$= H_m^\circ(\text{Fe(fcc)}, T) - H_m^\circ(\text{Fe(bcc)}, 298.15 \text{ K}) \qquad (9.2)$$

since Fe(bcc) is the stable form of Fe at 298.15 K. It can be seen in Figure 9.1a that $H_m^\circ(T)$ goes through zero at 298.15 K for the bcc form but not for Fe(fcc) or Fe(l).

In the case of $S_m^\circ(T)$ the Planck postulate is used for all crystalline substances so that we may write

$$S_m^\circ(\text{Fe(fcc)}, T) = S_m^\circ(\text{Fe(fcc)}, 298.15 \text{ K})$$
$$+ [S_m^\circ(\text{Fe(fcc)}, T) - S_m^\circ(\text{Fe(fcc)}, 298.15 \text{ K})] \qquad (9.3)$$

Materials Thermodynamics. By Y. Austin Chang and W. Alan Oates
Copyright 2010 John Wiley & Sons, Inc.

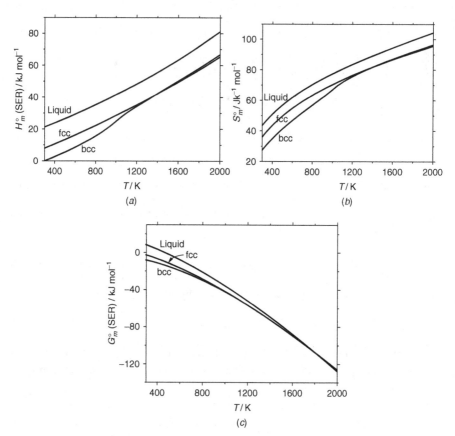

Figure 9.1 Values of standard state properties (using SER) for pure Fe in bcc, fcc, and liquid forms: (a) $H_m^\circ(\text{SER}, T)$; (b) $S_m^\circ(T)$; (c) $G_m^\circ(\text{SER}, T)$.

The values at 298.15 K are the standard entropies, none of which are zero. This can be seen in Figure 9.1b for the three structural forms of Fe. The standard Gibbs energy G_m° for Fe(fcc) is, as usual,

$$G_m^\circ(\text{Fe(fcc)},\text{SER}, T)$$
$$= H_m^\circ(\text{Fe(fcc)},\text{SER}, T) - T S_m^\circ(\text{Fe(fcc)}, T) \qquad (9.4)$$

and, as can be seen in Figure 9.1c, it does not go through zero at 298.15 K for any of the structural forms. It can also be seen that a definite curvature exists in the G_m°-versus-T curves, particularly at lower temperatures. This is handled by a polynomial representation, as was discussed in Chapter 2.

By using the SER, standard Gibbs energies and standard chemical potentials may be used as though they are *absolute* values.

(C) *Introduce Appropriate Solution-Phase Model*   For a component i in the α-phase at constant T, we can write

$$\mu_i^\alpha = \mu_i^{\circ\alpha}(\text{SER}) + \Delta_{\text{mix}}\mu_i^{\text{id},\alpha} + \mu_i^{\text{E},\alpha} \qquad (9.5)$$

and since $\mu_i^{\circ\alpha}(\text{SER}) = G_m^{\circ,\alpha}(i, \text{SER})$, we obtain the following equation for the chemical potential of a component in a solution phase:

$$\boxed{\mu_i^\alpha = G_m^{\circ,\alpha}(i, \text{SER}) + \Delta_{\text{mix}}\mu_i^{\text{id},\alpha} + \mu_i^{\text{E},\alpha}} \qquad (9.6)$$

Combination of the equilibrium conditions given in (9.1) with (9.6) for the two-phase equilibrium between α- and β-phases in a binary alloy (A,B) at constant T can now be written as

$$\mu_A^{\circ,\alpha}(\text{SER}) + \Delta_{\text{mix}}\mu_A^{\text{id},\alpha} + \mu_A^{\text{E},\alpha} = \mu_A^{\circ,\beta}(\text{SER}) + \Delta_{\text{mix}}\mu_A^{\text{id},\beta} + \mu_A^{\text{E},\beta} \quad (9.7)$$

$$\mu_B^{\circ,\alpha}(\text{SER}) + \Delta_{\text{mix}}\mu_B^{\text{id},\alpha} + \mu_B^{\text{E},\alpha} = \mu_B^{\circ,\beta}(\text{SER}) + \Delta_{\text{mix}}\mu_B^{\text{id},\beta} + \mu_B^{\text{E},\beta} \quad (9.8)$$

In order to proceed further, an appropriate modeling equation is required for the excess chemical potentials in the different solution phases. The most common way of representing the experimental results for liquid and substitutional solid phases is with a Redlich–Kister polynomial. For a binary system (A,B):

$$\mu_A^{\text{E}} = x_B^2 \sum_{j=0}^{n} L_j (x_A - x_B)^{j-1}[(2j+2)x_A - 1] \qquad (9.9)$$

$$\mu_B^{\text{E}} = x_A^2 \sum_{j=0}^{n} L_j (x_A - x_B)^{j-1}[1 - x_B(2j+2)] \qquad (9.10)$$

Example 9.1 Miscibility Gap Formation in Al–Zn System

We wish to calculate the equilibrium phase compositions and phase fractions for an alloy of overall composition $Al_{0.7}Zn_{0.3}$ at 500 K assuming that only the fcc phase has to be considered.

The following parameters are given. Note that the G_m° values for both elements have been linearized as a function of temperature to assist in simplifying the calculation:

$$G_m^\circ(\text{Al(fcc)}, \text{SER}) \approx 4627.33 - 41.18T \text{ J mol}^{-1}$$

$$G_m^\circ(\text{Zn(fcc)}, \text{SER}) \approx 7700.09 - 56.37T \text{ J mol}^{-1}$$

$$G_m^{\text{E}}(\text{fcc}) = x_{Zn}(1 - x_{Zn})[L_0 + L_1(1 - 2x_{Zn}) + L_2(1 - 2x_{Zn})^2]$$

$$L_0 = 6656.0 + 1.615T \text{ J mol}^{-1}$$

$$L_1 = 6793.0 - 4.982T \text{ J mol}^{-1}$$

$$L_2 = -5352.0 + 7.261T \text{ J mol}^{-1}$$

While Al(fcc) is the stable form of Al at 500 K, the stable form of Zn at this temperature is hcp. The above equation for $G_m^\circ(\text{Zn})$ thus refers to metastable Zn(fcc).

Miscibility gap formation represents one of the simplest type of phase diagrams since there is only one phase description to consider—at high temperatures a single phase exists over the whole composition range but at lower temperatures the single phase splits into two coexisting phases. This requires that $\Delta_{\text{mix}} H_m > 0$, but it should be remembered that the logarithmic nature of the mixing entropy term (and hence of $-T \Delta_{\text{mix}} S_m$) gives an infinite slope in the infinitely dilute solution, whereas $\Delta_{\text{mix}} H_m$ has a finite slope. This difference in asymptotic behavior ensures that there is always a minimum in $\Delta_{\text{mix}} G_m$ in dilute solution all the way down to 0 K.

A good example of a miscibility gap is found in fcc (Al,Zn) alloys. The stable phase diagram for the system is shown in Figure 9.2. Also shown (dashed) is the phase diagram which results when the liquid and hcp phases are suppressed to leave only the fcc phase to consider. Part of this fcc phase diagram is seen to represent a local and not a global equilibrium; that is, part is a metastable phase diagram. At 500 K, for example, the fcc phase equilibrium is metastable with respect to the fcc–hcp phase equilibrium.

Figure 9.3 shows $\Delta_{\text{mix}} G_m$ for the fcc alloys at various temperatures, both above and below the critical temperature of 622.3 K. It can be seen that the $\Delta_{\text{mix}} G_m$ versus x_{Zn} curve at the highest temperature is convex everywhere, indicating that the single phase is stable over the entire composition range. At the

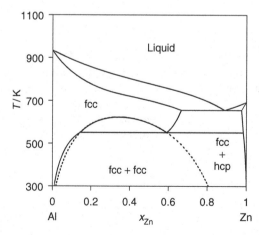

Figure 9.2 Stable Al–Zn phase diagram with metastable fcc phase diagram superimposed.

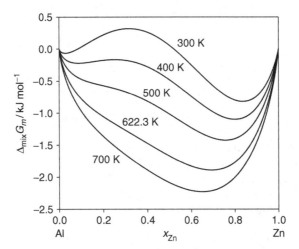

Figure 9.3 $\Delta_{\mathrm{mix}}G_m$ for fcc (Al,Zn) alloys at several temperatures.

three lowest temperatures, however, the $\Delta_{\mathrm{mix}}G_m$-versus-x_B curve exhibits two minima and one maximum, due to the fact that the $-T\,\Delta_{\mathrm{mix}}S_m$ term becomes of lesser importance with decreasing temperature.

The relevant properties for Al required in the calculation can be written as

$$\mu_{\mathrm{Al}}^{\circ,\alpha}(\mathrm{SER},\,500\ \mathrm{K}) = G_m^{\circ}(\mathrm{Al(fcc)},\,\mathrm{SER},\,500\ \mathrm{K}) = -15962.6\ \mathrm{J\ mol}^{-1}$$

$$\mu_{\mathrm{Al}}^{\mathrm{E},\alpha} = (x_{\mathrm{Zn}}^{\alpha})^2[L_0 + L_1(3 - 4x_{\mathrm{Zn}}^{\alpha}) + L_2(1 - 2x_{\mathrm{Zn}}^{\alpha})(5 - 6x_{\mathrm{Zn}}^{\alpha})]$$

There are analogous equations for Zn.

When these equations for both components in both phases are inserted into (9.7) and (9.8), the following solution is obtained:

$$x_{\mathrm{Zn}}^{\alpha} = 0.101 \qquad x_{\mathrm{Zn}}^{\alpha'} = 0.661 \qquad f^{\alpha'} = 0.356$$

Example 9.2 Solution/Solution-Phase Equilibrium in Fe–Mo System
We wish to calculate the equilibrium phase compositions of the liquid and bcc phases and their phase fractions for an $\mathrm{Fe}_{0.4}\mathrm{Mo}_{0.6}$ alloy at 2000 K and are required to provide a G_m–x plot which illustrates the results from the calculation.

The following parameters are given:

$$G_m^{\circ}(\mathrm{Fe(l)},\,SER) \approx 69714 - 98.38T\ \mathrm{J\ mol}^{-1}$$

$$G_m^{\circ}(\mathrm{Mo(l)},\,\mathrm{SER}) \approx 83383 - 91.98T\ \mathrm{J\ mol}^{-1}$$

$$G_m^{\circ}(\mathrm{Fe(bcc)},\,\mathrm{SER}) \approx 56097 - 90.85T\ \mathrm{J\ mol}^{-1}$$

$$G_m^{\circ}(\mathrm{Mo(bcc)},\,\mathrm{SER}) \approx 41690 - 77.38T\ \mathrm{J\ mol}^{-1}$$

$$G_m^E(l) = x_{Mo}(1 - x_{Mo})[L_0 + L_1(1 - 2x_{Mo})]$$

$$L_0(l) = -6973 - 0.37T \text{ J mol}^{-1}$$

$$L_1(l) = -9424 + 4.502T \text{ J mol}^{-1}$$

$$L_0(bcc) = 36818 - 9.141T \text{ J mol}^{-1}$$

$$L_1(bcc) = -362 - 5.724T \text{ J mol}^{-1}$$

The phase diagram for the Fe–Mo system is shown in Figure 9.4 with the tie-line relevant to the calculation shown dashed. The Gbbs energy curver for the two phases at the same temperature are shown in figure 9.5.

When the equations for both components in both phases are inserted into (9.7) and (9.8), the following solution is obtained.

$$x_{Mo}^{bcc} = 0.8315 \qquad x_{Mo}^{l} = 0.4784 \qquad f^l = 0.6566$$

Example 9.3 Solution/Stoichiometric Compound Phase Equilibrium in Mg–Si System
We wish to calculate:

(i) The liquid-phase composition and fraction at 1273 K for the alloy $Mg_{0.6}Si_{0.4}$ in equilibrium with the stoichiometric compound Mg_2Si
(ii) The activity of Mg relative to Mg(l) in these two phases
(iii) A G_m-x plot which illustrates the results from the calculation

The following parameters are given:

$$G_m^\circ(Mg(l), SER) \approx 35004 - 82.01T \text{ J mol}^{-1}$$

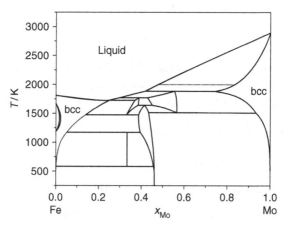

Figure 9.4 Phase diagram for Fe–Mo system. The tie-line at 2000 K for the equilibrium between the liquid and bcc phases is shown dashed.

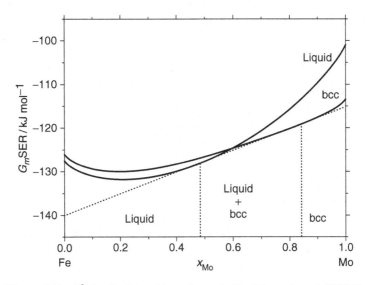

Figure 9.5 G_m° for liquid and bcc phases in Fe–Mo system at 2000 K.

$$G_m^\circ(\text{Si(l)}, \text{SER}) \approx 72406 - 82.103T \ \text{J mol}^{-1}$$

$$G_m^\circ(\text{Mg}_{0.667}\text{Si}_{0.333}(\text{s}), \text{SER}) = -2556.4 - 58.017T \ \text{J mol}^{-1}$$

$$G_m^{\text{E}} = x_{Si}(1 - x_{Si})[L_0 + L_1(1 - 2x_{Si})]$$

$$L_0 = -47052.64 \ \text{J mol}^{-1}$$

$$L_1 = -2385.71 \ \text{J mol}^{-1}$$

The Mg–Si phase diagram shown in Figure 9.6, calculated from the above approximated data, is an example of where a nearly stoichiometric compound is formed as a result of the strong chemical interaction between the components.

The maximum in the liquid at the Mg_2Si composition is known as a *congruent point*.

A rapid change in the component activities takes place in these "stoichiometric" compounds. This can be appreciated in Figure 9.7, which shows the variation of a_{Mg} [reference state Mg(l)] with composition over the whole composition range at 1273 K and can be understood while referring to the phase diagram in Figure 9.6 and the Gbbs energy curves shown in Figure 9.8.

The activity of Mg in the liquid phase first decreases gradually with increasing x_{Si} up to the solubility limit and then remains constant within the two-phase field until the stoichiometric composition is reached. It then decreases abruptly to a much lower value. The activity then remains constant within the two-phase field, decreases again as we pass through the liquid phase until it reaches (almost) zero, when it becomes saturated with almost pure Si(s) at higher concentrations of Si.

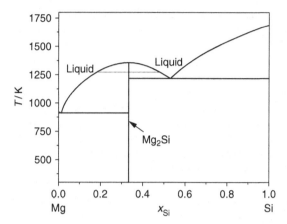

Figure 9.6 The Mg–Si phase diagram together with tie-line for equilibrium between Mg$_2$Si and liquid phase at 1273 K.

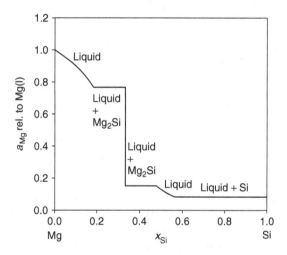

Figure 9.7 The a_{Mg} relative to Mg(l) at 1273 K in Mg–Si system.

There is only one unknown, the liquid-phase composition, and we can use (8.20), which, for this particular case, reads

$$0.667\mu_{Mg}^{l} + 0.333\mu_{Si}^{l} = G_m(\text{Mg}_2\text{Si}) \tag{9.11}$$

As is apparent from the phase diagram there are two solutions for the liquid-phase composition, one on each side of the stoichiometric compound. The two solutions can be obtained by appropriate selection of the starting values used in the equation solver. In the question, we are only required to find the solution on the Si-rich side of the compound.

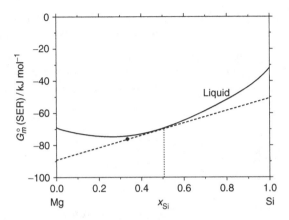

Figure 9.8   The $G_m^{\circ}(\mathrm{SER})$ for liquid and $\mathrm{Mg_2Si}$ phases at 1273 K.

When (9.11) is used in conjunction with (9.7) and (9.8), the following solution is obtained:

$$x_{Si}^l = 0.5063 \qquad f^{Mg_2Si} = 0.615$$

EXERCISES

9.1 Confirm the solutions for Examples 9.1–9.3 discussed in the text.

9.2 Components A and B form a regular solution in the solid state with a fcc structure designated by α for which $\Delta_{mix}H(\alpha) = 20{,}000 x_A x_B$ J mol^{-1}. Calculate the composition of α' and α'' in equilibrium and the chemical spinodal compositions at 1000 K.

9.3 Derive an expression which indicates the factors determining the maximum width of the liquid/solid two-phase region when both liquid and solid solutions are ideal.

9.4 Using the following data, calculate the Ag–Au phase diagram assuming that both the liquid and solid solutions are ideal. Compare the calculated diagram with the real one (see a reference book i.e., *Metals Handbook*).

	Ag	Au
T_{fus}°/K	1234	1336.15
$\Delta_{fus}S^{\circ}/$ J mol^{-1} K^{-1}	9.16	9.39

For semiconductors such as Si and Ge, the entropy of fusion is about 30.1 J K^{-1} mol^{-1}. Assume the entropy of fusion for Ag and Au to be 30.1 J K^{-1}

mol^{-1}. Recalculate the phase diagram for this hypothetical binary system, again assuming that the liquid and solid solutions behave ideally.

9.5 Using the data given below, calculate the Cu–Ni phase diagram assuming ideal solution behavior for both the liquid and solid phases. Compare the calculated diagram with the real one (see a reference book).

	Cu	Ni
T_{fus}°/K	1,356.55	1,726
$\Delta_{fus} H^{\circ}/$ J mol^{-1}	13,055	17,472

9.6 Figure 9.9 shows the phase diagram for the Ag–Pt system. Draw schematically the Gibbs energy curves versus composition diagrams for the Ag–Pt system at $T = 1300°C$, $1100°C$, and $900°C$.

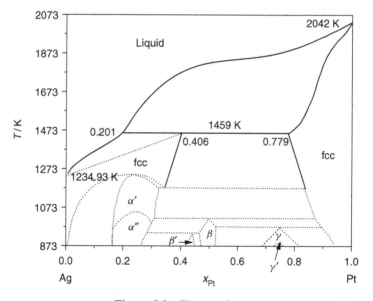

Figure 9.9 The Ag–Pt phase diagram.

9.7 Data for the pure elements and solution phases for the liquid and fcc phases in the (Al,Zn) system are given in the text.

   Carry out the following calculations *at 750 K* and present graphs summarizing the results from each calculation:

(a) Using the SER for both Al and Zn, plot G_m for the two phases. On the diagram draw the common tangent and obtain the equilibrium compositions of the two phases.

(b) Calculate x_B^α, x_B^β, f^β at $x_B = 0.3$, $T = 400$ K.

(c) Generate plots of another function which enables you to obtain these compositions more accurately than can be obtained from the common tangent to G.

(d) Repeat (a) and (b) using Al(l) and Zn(l) as the reference states and confirm that you get the same answer for the equilibrium compositions.

(e) Repeat (a) and (b) using Al(fcc) and Zn(fcc) as reference states for the fcc phase and Al(l) and Zn(l) as reference states for the liquid phase. Why do you now get a different answer for the two-phase equilibrium?

(f) Repeat (a)–(d) but obtain plots of μ_{Zn} for both phases. Label the meaning of the value of each intercept on the pure Zn axis, for example, μ_{Zn}°(fcc, 750 K)$-\mu_{Zn}^\circ$(l, 750 K).

9.8 For the temperature range 1000–1500 K, the activity coefficient of zinc in liquid brass can be expressed by the equation

$$RT \log_e \gamma_{Zn} = -19{,}245x_{Cu}^2 \, \text{J mol}^{-1}$$

Calculate the partial pressure of zinc, p_{Zn}, over a solution of 60 at % copper and 40 at % zinc at 1500 K.

10 Binary Phase Diagrams: Temperature–Composition Diagrams

We have indicated previously that, for unary systems, two-dimensional phase diagrams may be classified as follows:

Field–field phase diagrams
Field–density phase diagrams
Density–density phase diagrams

Field variables are intrinsically intensive quantities like p, T, μ_i. Density variables are intensive quantities formed from extensive quantities by division by either mass or amount and are referred to as specific quantities.

This same classification applies to phase diagrams for binary systems but, because of the additional component, there is an additional independent variable. It now requires three independent variables to specify the properties of a binary single phase. This means that, in order to be able to plot properties or phase diagrams in two dimensions, it is necessary to fix one of the independent variables. In MS&E, phase diagrams are invariably displayed at a fixed total pressure of 1 bar.

For binary systems, commonly met examples of the three types of plots are $T - \mu_i$, $T - x_i$, and $H_m - x_i$ diagrams, respectively. Schematic examples of these three are shown in Figure 10.1.

Just as in the case of unary systems, the topology of the three types of phase diagram differs. The three-phase invariant equilibrium is a point in field–field phase diagrams, a line in field–density phase diagrams, and an area in density–density phase diagrams.

As will be discussed in the next section, arbitrary plots of two properties do not necessarily result in a true phase diagram. There are constraints on the selection of the variables which can be plotted if a true phase diagram is to be obtained.

Materials Thermodynamics. By Y. Austin Chang and W. Alan Oates
Copyright 2010 John Wiley & Sons, Inc.

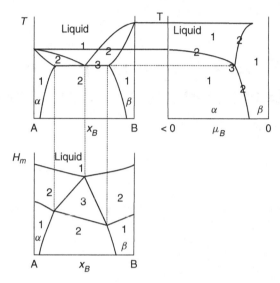

Figure 10.1 Examples of field–density, field–field, and density–density binary phase diagrams. The numerals indicate the number of phases present in that region of the diagram.

10.1   TRUE PHASE DIAGRAMS

Each thermodynamic potential has its *natural variables*, which are apparent from the appropriate Gibbs equation. For example, from $dU = T\,dS - p\,dV$, the natural variables of U are S and V. If the potential is known as a function of its natural variables, then all the thermodynamic properties of the system are fixed. Thus, for a multicomponent system, knowledge of $G = G(p, T, n_i)$ fixes all the other thermodynamic properties but, for example, knowing $G = G(V, T, n_i)$ does not.

Every extensive thermodynamic function has an intensive thermodynamic potential which is its *conjugate variable*. These pairs come from the definition of work in its widest, thermodynamic, sense. As can be seen from the fundamental equation

$$dU = T\,dS - p\,dV + \sum_i \mu_i\,dn_i \tag{10.1}$$

T and S are conjugate variables, as are $-p$ and V, and as are μ_i and n_i.

A plot of any pair of conjugate variables does not result in a true phase diagram, by which we mean one where all points on the diagram give a unique answer to the phase constitution at that point. Plots of conjugate variables simply lead to a property diagram, not a phase diagram.

A multitude of thermodynamically valid equations similar to (10.1) are possible. As an example, consider the constant-pressure Gibbs–Duhem equation for a binary system:

$$0 = S\,dT + n_A\,d\mu_A + n_B\,d\mu_B \tag{10.2}$$

We may rewrite this equation in two different ways, first, dividing by n_B and, second, dividing by $n_A + n_B$:

$$-d\mu_B = \frac{S}{n_B}dT + \frac{n_A}{n_B}d\mu_A \tag{10.3}$$

$$-d\mu_B = S_m\,dT + x_A\,d(\mu_A - \mu_B) \tag{10.4}$$

Using the rule that the plotting of conjugate variables does not result in a true phase diagram, we can see from (10.3) that, at constant total p, the following types of true phase diagrams may be drawn:

> Field–field: T versus μ_A
>
> Field–density: μ_A versus $\dfrac{S}{n_B}$ and T versus $\dfrac{n_A}{n_B}$
>
> Density–density : $\dfrac{S}{n_B}$ versus $\dfrac{n_A}{n_B}$

Plots of S/n_B versus T or of n_A/n_B versus μ_A do not give true phase diagrams.

If the molar quantities in (10.4) are used, the following types of true phase diagrams, at constant p, are indicated:

> Field–field: T versus $\mu_A - \mu_B$
>
> Field–density: $\mu_A - \mu_B$ versus S_m and T versus x_B
>
> Density–density: S_m versus x_B

Plots of S_m versus T or of x_A versus $\mu_A - \mu_B$, on the other hand, are not true phase diagrams.

Depending on the thermodynamic equation used, other sets of conjugate variables are possible and this leads to other pairs of variables which can be plotted to give a true phase diagram and those which cannot. A field–field diagram will always be a true phase diagram because the field variables are not conjugate. Similarly, density–density diagrams are also always true phase diagrams, with the proviso that the same mass or amount units are used for both functions. It is with field–density diagrams that more care has to be exercised. This is illustrated in Figure 10.2 where some different choices of variables for the simple eutectic system formed by (Pb,Sn) alloys, based on (10.3) and (10.4), have been selected. It can be seen that three of these diagrams are true phase diagrams but that (Fig. 10.2c) is not. The shaded area in this field–density diagram indicates a region where any point does not give a unique answer for the phase constitution at that point.

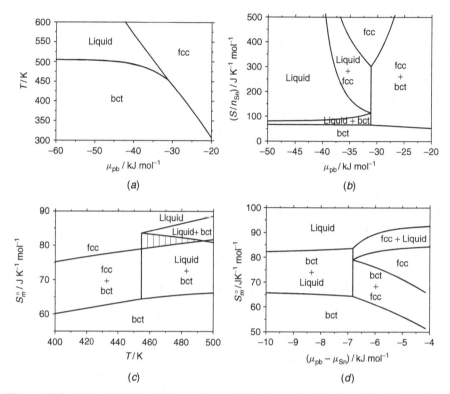

Figure 10.2 Plots of different thermodynamic functions for Pb–Sn system illustrating what are and what are not true phase diagrams: (*a*) field–field and (*b*) density–field true phase diagrams; (*c*) density–field diagram of conjugate variables; (*d*) density–field true phase diagram. The SER reference states have been used.

Many different types of phase diagram have a practical value. Plots of $H_m - x_i$, for example, are useful in heat balance calculations. The type of most significance in MS&E is the $T-x_i$ diagram and we now discuss some aspects of these in more detail.

10.2 $T-x_i$ PHASE DIAGRAMS FOR STRICTLY REGULAR SOLUTIONS

Application of (9.7) and (9.8) to a liquid and solid (α) phases with the assumption that both solution phases are strictly regular gives, at constant p, T,

$$\mu_A^l = \mu_A^\alpha \tag{10.5}$$

$$\mu_A^{\circ,\alpha} + \Delta\mu_A^{id,\alpha} + \mu_A^{E,\alpha} = \mu_A^{\circ,l} + \Delta\mu_A^{id,l} + \mu_A^{E,l}$$
$$- (\mu_A^{\circ,l} - \mu_A^{\circ,\alpha}) \tag{10.6}$$

The last term in (10.6) can be expressed as

$$(\mu_A^{\circ,1} - \mu_A^{\circ,\alpha}) = \Delta_{\text{fus}} H_m^{\circ}(A) - T \Delta_{\text{fus}} S_m^{\circ}(A) \tag{10.7}$$

$$\approx [T_{\text{fus}}^{\circ}(A) - T] \Delta_{\text{fus}} S_m^{\circ}(A) \tag{10.8}$$

so that we obtain the working equation for component A:

$$RT \log_e(1 - x_B^1) + L_0^1(x_B^1)^2 = RT \log_e(1 - x_B^{\alpha}) + L_0^{\alpha}(x_B^{\alpha})^2$$
$$- [(T_{\text{fus}}^{\circ}(A) - T) \Delta_{\text{fus}} S_m^{\circ}(A)] \tag{10.9}$$

$$RT \log_e \left(\frac{x_A^1}{x_A^{\alpha}} \right) = [L_0^{\alpha}(x_B^{\alpha})^2 - L_0^1(x_B^1)^2]$$
$$- [(T_{\text{fus}}^{\circ}(A) - T) \Delta_{\text{fus}} S^{\circ}(A)] \tag{10.10}$$

with a similar equation for the other component, B. The two equations can be solved to calculate binary phase diagrams for strictly regular solutions.

From (10.10) and the analogous equation for B, we see that the location of phase boundaries depends on:

(i) $T_{\text{fus}}^{\circ}(A)$ and $T_{\text{fus}}^{\circ}(B)$
(ii) $\Delta_{\text{fus}} S_m^{\circ}(A)$ and $\Delta_{\text{fus}} S^{\circ}(B)$
(iii) L_0^1 and L_0^{α}

As is illustrated below, variation of these factors can have a profound effect on the phase diagram topology for these strictly regular solutions. For most metals, $\Delta_{\text{fus}} S_m^{\circ} \approx 10$ J K^{-1} mol^{-1} while it may be much higher than this for semiconductors. For the other pure component properties we have selected two sets of parameters, one set where there is a small difference in T_{fus}° of the two components and the other set with a large difference:

$T_{\text{fus}}^{\circ}(A) = 1000$ K	$T_{\text{fus}}^{\circ}(A) = 500$ K
$T_{\text{fus}}^{\circ}(B) = 1200$ K	$T_{\text{fus}}^{\circ}(B) = 3000$ K
$\Delta_{\text{fus}} S_m^{\circ}(A) = 10$ J mol^{-1} K^{-1}	$\Delta_{\text{fus}} S_m^{\circ}(A) = 10$ J mol^{-1}
$\Delta_{\text{fus}} S_m^{\circ}(B) = 10$ J mol^{-1} K^{-1}	$\Delta_{\text{fus}} S_m^{\circ}(B) = 10$ J mol^{-1}K^{-1}

In Figure 10.3, the phase diagrams on the left have been calculated using the first set of parameters, with a small difference in fusion points of the components; those on the right are obtained when large differences in the fusion points are selected.

The effect of varying the solution parameters of the two phases are apparent in the figure.

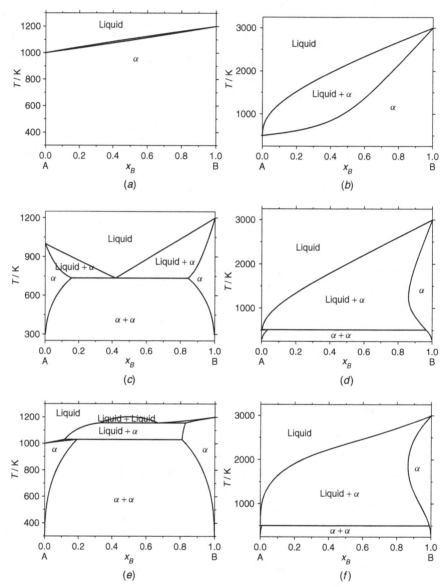

Figure 10.3 Phase diagrams involving two phases, both of which are strictly regular solutions. There is a 200-K difference in melting points for those in the left column, a 2500-K difference for those in the right column: (a, b) $L_0^l = 0$, $L_0^\alpha = 0$; (c, d) $L_0^l = 0$; $L_0^\alpha = +15$ kJ mol^{-1}; (e, f) $L_0^l = +20$ kJ mol^{-1}, $L_0^\alpha = +20$ kJ mol^{-1}.

10.2.1   Some General Observations

It is clear from these phase diagrams for strictly regular solutions that a considerable variety can be found by changing just a few model parameters. An examination of diagrams in Figures 10.3 and 10.4 enables us to reach some general observations as to when we might expect to see different types of simple phase diagrams.

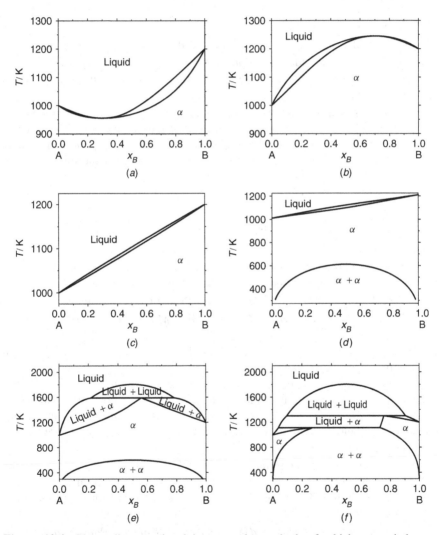

Figure 10.4 Phase diagrams involving two phases, both of which are strictly regular solutions. All refer to systems with a 200-K difference in melting points: (a) $L_0^l = -5$ kJ mol^{-1}, $L_0^\alpha = 0$; (b) $L_0^l = 0$, $L_0^\alpha = -5$ kJ mol^{-1}; (c) $L_0^l = -10$ kJ mol^{-1}, $L_0^\alpha = -10$ kJ mol^{-1}; (d) $L_0^l = +10$ kJ mol^{-1}, $L_0^\alpha = +10$ kJ mol^{-1}; (e) $L_0^l = +30$ kJ mol^{-1}, $L_0^\alpha = +10$ kJ mol^{-1}; (f) $L_0^l = +30$ kJ mol^{-1}, $L_0^\alpha = +20$ kJ mol^{-1}.

(i) *Complete Solubility Phase Diagrams*   Complete solubility results when both phases are ideal, that is, $L_0^l = L_0^\alpha = 0$. But we can have a complete solubility phase diagram even when the solutions are not ideal. Such diagrams result when $L_0^l \approx L_0^\alpha$. Real alloy examples include the Ag–Au, Ag–Pd, Cd–Mg, and Co–Ni systems.

When both L_0^l and L_0^α are negative, the phase diagram is similar to that for the case $L_0^l = L_0^\alpha = 0$, but when $L_0^l \approx L_0^\alpha \gg 0$, miscibility gaps are formed at lower temperatures. A real alloy example of this is found in the Au–Ni system, where $L_0^\alpha \gg 0$.

Other features to note are:

> (i) When $L_0^l \approx L_0^\alpha \gg 0$, the width of the melting region becomes larger than for the ideal solution case. Conversely, when $L_0^l \approx L_0^\alpha < 0$, the width of the melting region is narrower than for the ideal solution case.

> (ii) The width of the two-phase region also depends on the relative melting points and entropies of fusion of the components. The freezing ranges for semiconductor alloys (e.g., Ge–Si) are particularly large, due primarily to the large entropies of fusion for the two elemental semiconductors.

(ii) *Peritectic Phase Diagrams*   Peritectic phase diagrams are more likely to occur when:

> (a) There is a large difference between $T_{fus}^\circ(A)$ and $T_{fus}^\circ(B)$.
> (b) $L_0^l \approx L_0^\alpha > 0$.
> (c) The critical temperature T_c for the (metastable) miscibility gap of the solid phase intercepts the (metastable) solidus curve.

> Real alloy examples include the Ag–Pt and Co–Cu systems.

(iii) *Phase Diagrams with Maximum and Minimum Congruent Melting Points*   A maximum occurs when the solid phase is more stable than the liquid phase and vice versa but when $L_0^l \approx L_0^\alpha$ and both are small. If they are too high and positive, miscibility gaps will occur.

When $L_0^l < L_0^\alpha$ and L_0^l is not too large, a maximum occurs in the liquid/solid. An example of a congruent maximum in a real alloy system occurs in the Pb–Tl system. When $L_0^l > L_0^\alpha$ and L_0^α is not too large, a minimum occurs in the liquid/solid. Examples of real alloy systems showing congruent minima are Co–Pd and Cr–Mo.

(iv) *Eutectic Phase Diagrams*   When $L_0^\alpha > > 0$ and $L_0^l \leq 0$ and the difference in melting points of the two components is not too large, a eutectic phase diagram is found. Cu–Ag is a real alloy system with this type of phase diagram.

(v) *Monotectic, Syntectic, and Monotectic + Syntectic Phase Diagrams*   Miscibility gaps form in the liquid phase when $L_0^l \gg 0$. Pb–Zn is a real alloy system which shows a liquid-phase miscibility gap.

10.2.2   More on Miscibility Gaps

By examining the compositional dependencies of G and its derivatives, we can obtain the conditions which pertain at the critical temperature. We illustrate by reference to the Al–Zn system, which was discussed as Example 9.1, the stable phase diagram for the system having been given in Figure 9.2. The stable miscibility gap exists over a fairly small temperature range. Its metastable extension to lower temperatures is shown in Figure 10.5a. The critical point occurs at approximately 622.3 K. The first three derivatives of the Gibbs energy of mixing with respect to x_{Zn} for the fcc (Al,Zn) phase at this temperature are shown in Figures 10.5b–10.5d.

It can be seen that, at the critical point, $d^2G/dx_B^2 = d^3G/dx_B^3 = 0$ but that $dG/dx_B \neq 0$ (we have abbreviated $\Delta_{mix}G$ to G). We have selected the fcc (Al,Zn) example since it is not a regular solution. For the latter, the critical point occurs at $x = 0.5$, and this gives rise to a special case where the first derivative is also zero at that point. This is not a general condition.

The zero values of the second and third derivatives at the critical point are relevant to an apparent breakdown of the phase rule when applied to miscibility gap systems at the critical point. Clearly, $F = 0$ at this point since the

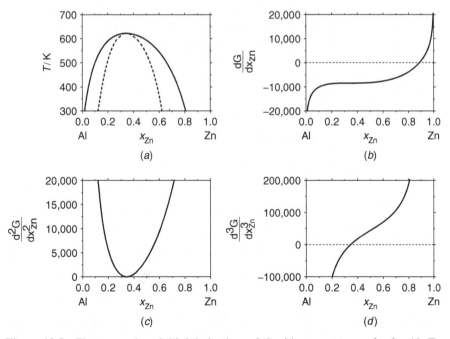

Figure 10.5   First, second, and third derivatives of G with respect to x_{Zn} for fcc Al–Zn at critical temperature of 622.29 K: (*a*) miscibilty gap in fcc phase; (*b*) first derivative of G; (*c*) second derivative of G; (*d*) third derivative of G. The critical composition occurs at $x_{Zn} = 0.347$.

temperature and composition are fixed (at constant total pressure). But, if we apply (8.26), we find that $F = 2 - 1 + 1 = 2$, which is incorrect. The reason is that there are other constraints on the μ_i at this point and we must apply (8.21). The extra constraints are those already discussed, namely, at the critical point, $d^2 G_m/dx_B^2 = d^3 G_m/dx_B^3 = 0$, so that, with the Gibbs–Duhem relation, $R_\mu = 3$. Applying (8.21), we see that $F = 1 \times 2 + 1 - 3 = 0$, as it should be.

10.2.3   The Chemical Spinodal

The dashed curve shown in Figure 10.5a is the locus of compositions where $d^2 G/dx_B^2 = 0$ and is called the *chemical spinodal* curve. The phase boundary for the two-phase stable equilibrium is referred to as the binodal curve.

The significance of the spinodal can be understood by referring to Figure 10.6. The points s_1 and s_2 are the spinodal points.

When a single-phase alloy with a composition between the spinodals is brought from a high temperature $(T > T_c)$ to a lower temperature within the miscibility gap, this alloy becomes thermodynamically unstable. The equilibrium system will consist of a two-phase mixture. There are, however, two quite different ways in which the phase decomposition can occur:

(a) Between the spinodal compositions, the $G - x_B$ curve is concave. In this case, the Gibbs energy of the homogenous single-phase alloy is always higher than that of any mixture of compositions (c and d, for example) which can form from a decomposition of this phase.

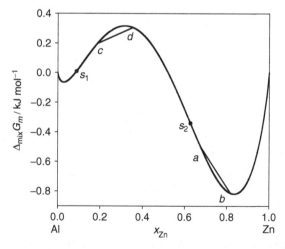

Figure 10.6   The $\Delta_{\mathrm{mix}} G_m^\circ$ for fcc (Al,Zn) at 300 K. At compositions inside the spinodal, any fluctuations in composition are unstable, whereas in the metastable region, between the binodal and the spinodal, fluctuations are stable.

(b) When a single-phase alloy with a composition between the miscibility gap and the nearby spinodal composition is brought to a $T < T_c$, the situation is quite different. Due to the convex nature of the $G - x_B$ curve in this composition range, there would be an increase of the Gibbs energy if a single-phase alloy decomposed locally to form any mixture of compositions (a and b, for example); that is, there is no thermodynamic driving force for decomposition by fluctuations when the composition falls between the spinodal and binodal. Decomposition in this case can only take place by nucleation of a phase in the other.

These two different mechanisms, spinodal decomposition and nucleation, are very important in the theory of phase transformations.

10.3   POLYMORPHISM

The Hf–Ta phase diagram shown in Figure 10.7 indicates that there is an allotropic transformation in Hf. The hcp form is stable at low temperatures with a transformation to bcc occurring at 1743°C. Tantalum is bcc at all temperatures in the solid state. It can be seen that the two bcc forms are completely miscible. But the difference in the two structures at low temperatures brings about a completely different topology to the phase diagram than would have been the case if only the high-T form of Hf existed. The method of calculation of the phase equilibrium boundaries between the Hf-rich hcp phase and the bcc phase is identical with that discussed above for the liquid/α equilibrium.

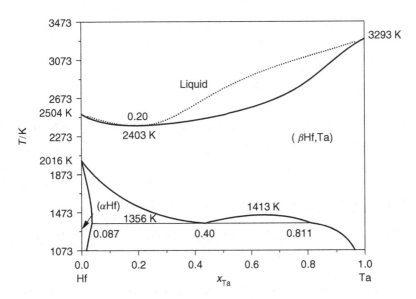

Figure 10.7   The Hf–Ta phase diagram.

EXERCISES

10.1 In the Mo–Ru phase diagram given in Figure 10.8, there exists a bcc (Mo, Ru) solid solution, a hcp (Ru, Mo) solid solution, an intermetallic phase denoted by σ, and a liquid phase. Suppose the σ-phase does not exist in this system. Estimate $\Delta G_{\text{bcc}}^{\text{hcp}}(\text{Mo})$ and $\Delta G_{\text{hcp}}^{\text{bcc}}(\text{Ru})$ at 1800°C on the basis of the solubility of Ru in bcc (Mo, Ru) and that of Mo in hcp (Ru, Mo), making any needed assumption to carry out this calculation.

   The solubility curves in Figure 10.8 were determined for the bcc + σ and σ + hcp two-phase fields. If σ does not form due to kinetic difficulties, we would then have a metastable equilibrium between bcc and hcp. Would the solubilities of Ru in bcc (Mo, Ru) and of Mo in hcp (Ru, Mo) decrease or increase for the metastable equilibrium?

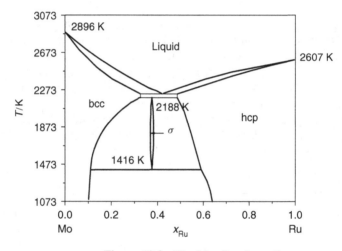

Figure 10.8 The Mo–Ru phase diagram.

10.2 Derive the critical point conditions for a subregular solution with the following expression for G^E:

$$G^E = x_A x_B [L_0 + L_1(x_a - x_B)] \tag{10.11}$$

10.3 G_m^E for fcc (Al,Zn) alloys may be represented by

$$G_m^E/\text{J mol}^{-1} = x_{\text{Al}}x_{\text{Zn}}[6656.0 + 1.615T) + (6793.0 - 4.982T)(x_{\text{Al}} - x_{\text{Zn}}) + (-5352.0 + 7.261T)(x_{\text{Al}} - x_{\text{Zn}})^2]$$

Calculate the miscibility gap and evaluate the critical point (approximately). At the critical temperature, plot $\Delta_{\text{mix}}G_m$ and the first, second, and third derivatives of $\Delta_{\text{mix}} G_m$ as a function of x_{Zn} and note which ones are zero at the critical composition.

10.4 If it is assumed that both the liquid and solid solution phases are ideal, a very simple phase diagram results in which the solid solution is stable down to 0 K. How does this tie in with your understanding of the Planck postulate?

11 Binary Phase Diagrams: Temperature–Chemical Potential Diagrams

To date, we have considered only temperature–composition phase diagrams, which, as we have seen, are actually a constant total pressure slice through a three-dimensional diagram in pressure–temperature–composition space. These phase diagrams are the most widely used two-dimensional phase diagrams for binary systems. There are good reasons for this:

(i) Examination of temperature–composition space is the normal way of experimentally obtaining phase equilibrium information.

(ii) Alloys are usually used in temperature–composition space in their production and applications.

In thermodynamic language, these two points mean that we usually carry out measurements on and use binary systems, especially alloys, in a closed-system sense.

This $T-x_i$ representation, however, is not always the most useful and it is certainly not the only way of graphically presenting information about phase equilibrium.

In chapter 10, we noted that, in a true phase diagram, any point on that diagram gives a unique answer as to the phase constitution represented at that point. We also noted that a field–field phase diagram is always a true phase diagram since two field variables can never be conjugate variables. The most useful of the field–field diagrams, whenever an open system is being considered (where composition is not fixed), is a plot of T versus μ_i at constant total pressure. When one or more of the components (or compounds of the components) are volatile at the temperatures of interest, composition may then become a dependent variable with an external chemical potential being the independent variable.

We first discuss some general points about these $T-\mu_i$ phase diagrams and will then discuss some examples. This same type of representation will be discussed again in Chapters 19–22, when we will refer to heterogeneous chemical equilibrium. For the moment, we concentrate on phase equilibrium.

Materials Thermodynamics. By Y. Austin Chang and W. Alan Oates

11.1   SOME GENERAL POINTS

In earlier chapters we placed emphasis on the value of the SER in carrying out equilibrium calculations. This reference state is not, of course, the only possible selection. It is not usually the reference state of choice in the presentation of $T-\mu_i$ phase diagrams. A more useful choice (but not a mandatory one) is to take the pure component at T, $p = 1$ bar as the reference state. By making this selection, the $T-\mu_i$ phase diagram is made to resemble the equivalent $T-x_i$ phase diagram (remember that there is a monotonic relation between chemical potential and composition in single-phase regions).

This selection of the reference state selection presents no ambiguity in the case of gases but, for condensed phases, it is necessary to decide whether to select the pure solid or the pure liquid component at all temperatures (and standard pressure) as the reference state. In the remainder of this chapter we will use the pure solid reference state so that, when we write μ_i, we mean $\mu_i(T, 1 \text{ bar}, x_i) - \mu_i^{\circ,s}(T, 1 \text{ bar}, x_i = 1)$. Since we will concentrate on $T-\mu_B$ phase diagrams for an A–B system, we are selecting $B(s)$ at the T of interest as the reference state.

Example 11.1 System with Two Ideal Solution Phases

As an introduction to $T-\mu_i$ diagrams, we will consider the simplest possible case where both the liquid and solid phases form ideal solutions. The $T-x_B$ phase diagram is shown in Figure 11.1. This phase diagram has been calculated on the assumption that $\Delta_{\text{fus}}S_m^{\circ} = 10 \text{ J K}^{-1} \text{ mol}^{-1}$ for both components. The chosen melting points are apparent on the phase diagram. It can be seen from Figure 11.1 that a liquid alloy with $x_B = 0.1$ is in equilibrium with an α phase alloy with $x_B = 0.374$ at 1432 K.

Having selected the $B(s)$ as the reference state, it is necessary to calculate a line on the $T-\mu_B$ diagram representing the properties of pure $B(l)$. This can be

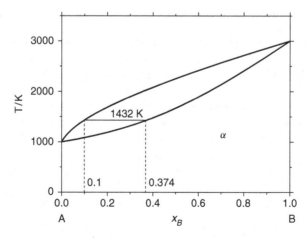

Figure 11.1   The $T-x_B$ phase diagram where both solid and liquid form ideal solutions.

obtained as follows:

$$\Delta_{\text{fus}}\mu_B^{\circ} = \mu_B^{\circ,\text{l}}(T) - \mu_B^{\circ,\text{s}}(T) = \Delta_{\text{fus}}G_m^{\circ}(B, T)$$
$$\simeq \langle\Delta_{\text{fus}}H_m^{\circ}(B)\rangle - T\langle\Delta_{\text{fus}}S_m^{\circ}(B)\rangle \qquad (11.1)$$

where, in the last equation, we have used the linear approximation for $\Delta_{\text{fus}}G_m^{\circ}$. The bracketed terms are average values for the quantities. This linear approximation is quite satisfactory when used in high-temperature calculations (above room temperature). The average values in (11.1) refer to values obtained from high-temperature results, even though they are used, but only for construction purposes, at lower temperatures, including 0 K. In this linear approximation, these construction lines are given by:

(i) At $T = 0$ K : $\Delta_{\text{fus}}\mu_B^{\circ}(0\ \text{K}) = \langle\Delta_{\text{fus}}H_m^{\circ}(B)\rangle$
(ii) At $T = T_{\text{fus}}^{\circ}$: $\Delta_{\text{fus}}\mu_B = 0$

so that it is possible to draw the line going through these two points on a T–μ_B diagram. This is illustrated in Figure 11.2. We can extend the calculations to alloys by putting lines on the T–μ_B phase diagram for different alloy compositions in both the solid and liquid phases. Since both the solid and liquid alloys have been assumed to form ideal solutions, the chemical potentials in the solid and liquid phases, relative to pure solid B as the reference state, are given by

$$\mu_B^{\text{s}}(x_B) - \mu_B^{\circ,\text{s}} = RT\log_e x_B^{\text{s}} \qquad (11.2)$$
$$\mu_B^{\text{l}}(x_B) - \mu_B^{\circ,\text{s}} = RT\log_e x_B^{\text{l}} + \Delta_{\text{fus}}G_m^{\circ}(B) \qquad (11.3)$$

These two equations give rise to composition fans on the T–μ_B diagram. These fans for the solid solution phase originate, at $T = 0$ K, according to (11.2)

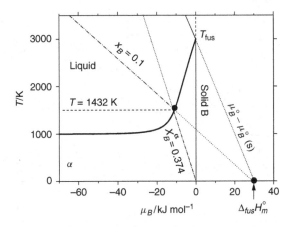

Figure 11.2 The T–μ_B phase diagram where both solid and liquid form ideal solutions.

at the origin, while those for the liquid phase originate, according to (11.3), at $T = 0$ K at the point $\Delta_{fus} H_m^\circ (B)$.

The $T-\mu_B$ phase diagram shown in Figure 11.2 is seen to be similar to the $T-x_B$ diagram given in Figure 11.1 but with the difference that the two-phase region in the $T-x_B$ diagram has contracted to a single curve in the $T-\mu_B$ diagram. The line for $x_B = 0.1$ for the liquid phase emanates from the point on the 0 K axis marked as $\Delta_{fus} H_m^\circ$. It can be seen to intersect with the composition line for $x_B = 0.374$ for the solid solution phase, which emanates from the origin of the figure, at the temperature of 1432 K.

It is clear that there is exactly the same information on the $T-\mu_B$ diagram as on the $T-x_B$ phase diagram. Note that we could show μ_B on the $T-x_B$ diagram so that then both diagrams would contain exactly the same information. Because the introduction of the μ_B lines on the $T-x_B$ phase diagram would lead to clutter, it is much more convenient to use the $T-\mu_B$ plot in those circumstances where μ_B is under external control.

It is also possible to include the gas-phase properties on both the $T-x_B$ and $T-\mu_B$ diagrams. Taking the same example as above and assuming that $\Delta_{vap} G_m^\circ = 340,000 + 85T$ J mol^{-1}, which gives a normal boiling point of 4000 K, we can insert curves for selected p_B on the $T-x_B$ phase diagram, as shown in Figure 11.3. The inclusion of gas-phase partial pressure fans on a $T-\mu_B$ diagram requires a very similar procedure to that used for composition fans for the liquid and solid phases in Figure 11.2. Phase equilibrium is governed by the relation $\mu_B^g = \mu_B^s$. In order to incorporate the gas-phase properties on the $T-\mu_B$ diagram, they must be referred to the pure solid B reference state.

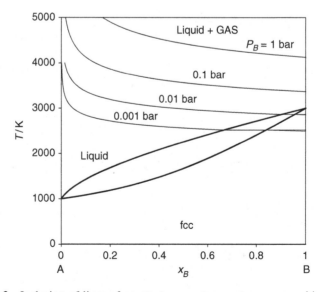

Figure 11.3   Inclusion of lines of constant p_B on temperature–composition diagram.

For the standard property difference between gas and solid, we may write (using the abbreviation sub for sublimation)

$$\mu_B^{\circ,g}(x_B) - \mu_B^{\circ,s} = \Delta_{sub}G_m^{\circ}(B, T) \tag{11.4a}$$

$$\simeq \langle \Delta_{sub}H_m^{\circ}(B) \rangle - T \langle \Delta_{sub}S_m^{\circ}(B) \rangle \tag{11.4b}$$

where we have again used the linear approximation. From (11.4) we see that, in the linear approximation, the construction lines are given by:

(i) At $T = 0$ K: $\mu_B^{\circ,g}(0 \text{ K}) - \mu_B^{\circ,g}(0 \text{ K}) = \langle \Delta_{sub}H_m^{\circ}(B) \rangle$

(ii) At $T = T_{sub}^{\circ}$: $\mu_B^{\circ,g}(T_{sub}^{\circ}) - \mu_B^{\circ,s}(T_{sub}^{\circ}) = 0$

If the gases are assumed perfect, we may use

$$\mu_B^g(p_B) - \mu_B^{\circ,g} = RT \log_e \left(\frac{p_B}{p_B^{\circ}} \right) \tag{11.5}$$

The chemical potential relative to pure solid B is then obtained by adding (11.4a) and (11.5):

$$\mu_B^g(p_B) - \mu_B^{\circ,s} = RT \log_e \left(\frac{p_B}{p_B^{\circ}} \right) + \Delta_{sub}G_m^{\circ}(B, T) \tag{11.6}$$

We can now superimpose the gas-phase information onto Figure 11.2 to yield a phase diagram from which we are able to read the liquid or solid compositions in equilibrium with a gas phase of fixed p_B at any temperature T. The resulting $T-\mu_B$ phase diagram is shown in Figure 11.4.

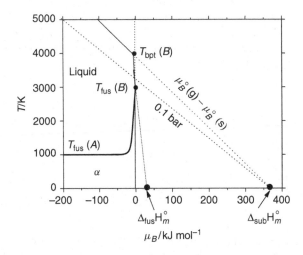

Figure 11.4 Inclusion of lines of constant p_B on $T-\mu_B$ phase diagram.

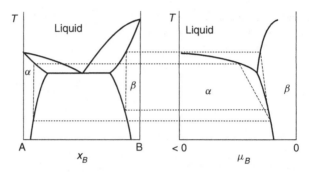

Figure 11.5 Example of interrelation between $T-x_B$ and $T-\mu_B$ phase diagrams. The inclusion of isoconcentration lines on the $T-\mu_B$ phase diagram is illustrated.

Example 11.2 Simple Eutectic System

Some practice is required in the interconversion of $T-x_B$ and $T-\mu_B$ phase diagrams. A slightly more complicated example is shown in Figures 11.5. The three-phase equilibrium line in the $T-x_B$ diagram has become a point in the $T-\mu_B$ diagram. The melting point of pure B shown in the $T-x_B$ diagram corresponds with $\mu_B = 0$ in the $T-\mu_B$ diagram, while the melting point of pure A corresponds with $\mu_B = -\infty$ in the $T-\mu_B$ diagram. The location of two isocomposition curves on the two diagrams is also apparent.

Example 11.3 More Complex System

Figure 11.6 shows how to convert from a $T-\mu_B$ diagram to a $T-x_B$ diagram in a more complex case. Three stable three-phase points are apparent in Figure 11.6a. The first thing to do is to focus on these invariant points and translate them to $T-x_B$ space as in Figure 11.6b—their relative position in the x_B direction will be expected to be the same as their relative positions in the μ_B direction. The two-phase boundaries may then be joined to give Figure 11.6c.

Note that, in Figure 11.6a, the extrapolations of the $\alpha\beta$, $\beta\gamma$, and $\gamma\alpha$ two-phase curves must intersect, as shown dotted, to give a single triple point for the metastable $\alpha\beta\gamma$ equilibrium.

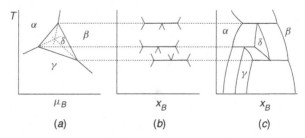

Figure 11.6 Example of interrelation between $T-x_B$ and $T-\mu_B$ phase diagrams. The way in which invariant points appear on both types of phase diagram is illustrated.

Example 11.4 Preparation of GaAs

In the making of a (Pb,Sn) solder, Pb and Sn can be weighed out in a molar ratio of about 1 : 3. The resulting alloy will act as a low-melting-point solder with little mushy freezing. If we are a little out in the weighing of the constituents, it will not make all that much difference to the engineering utility of the alloy.

This preparation of a solder may be contrasted with the very different situation for the preparation of GaAs, whose engineering applications depend on its electrical properties. According to the normal temperature–composition phase diagram shown in Figure 11.7a, GaAs is essentially stoichiometric. But, if we make one alloy with (0.4999 Ga, 0.5001 As) and one with (0.5001 Ga, 0.4999 As), then it is found that these two alloys have completely different engineering properties. This is because GaAs is marginally nonstoichiometric, as is apparent in a gross expansion of the composition scale of the temperature–composition phase diagram in the neighborhood of the stoichiometric composition (see Fig. 11.7b). An alloy composition cannot be guaranteed to this degree of precision if it is made up by weighing. Fortunately, however, it is possible to take advantage of

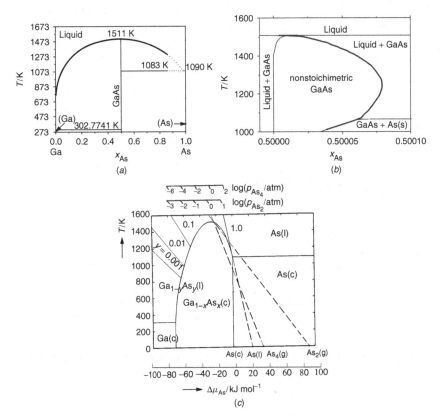

Figure 11.7   Phase diagrams for Ga–As system: (a) normal T–x_{As} and (b) expanded T–x_{As} phase diagrams; (c) T–μ_{As} phase diagram.

the fact that μ_{As} changes considerably over this very narrow composition range. In practical terms, orders-of-magnitude changes in the arsenic pressure can bring about minuscule, but extremely significant, changes in composition. This is clear in the $T-\mu_{As}$ phase diagram shown in Figure 11.7c.

It is quite clear that, to those interested in the production of single crystals of GaAs, a $T-x_i$ phase diagram is not as useful as a $T-\mu_i$ phase diagram.

As we will see in later chapters, open systems are frequently encountered when examining heterogeneous equilibria in chemically reacting systems. In these situations, $T-\mu_i$ diagrams are the more useful way for graphically presenting the results.

EXERCISES

11.1   Calculate $T-\mu_A$ phase diagrams from the $T-x_B$ phase diagrams shown in Figures 11.1 and 11.5.

11.2   Figure 11.8 shows a "double" Fe–C temperature–composition phase diagram. The heavy lines refer to the iron–graphite system and the dashed lines to the metastable iron–carbide system. To those interested in carburizing or decarburizing of Fe, a $T-\mu_C$ phase diagram is a more useful type of presentation. Use the $T-x_C$ double diagram to sketch the $T-\mu_C$ phase diagram using graphite as the reference state at all temperatures.

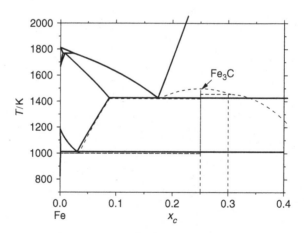

Figure 11.8   The $T-x_C$ stable and metastable phase diagrams for Fe–C system.

11.3 Sketch the $T - x_B$ phase diagram corresponding to the $T - \mu_B$ phase diagram shown in Figure 11.9. Include, with dashed lines, the location of the metastable $\alpha\beta\gamma$ invariant equilibrium and the associated univariant equilibria.

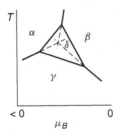

Figure 11.9 A $T - \mu_B$ phase diagram.

12 Phase Diagram Topology

As we have seen in the previous chapters, some of the constant-pressure $T - x_i$ phase diagrams for real binary alloys are fairly simple. Many others, on the other hand, seem quite complex but all are comprised of one-phase regions, two-phase regions, and horizontal lines connecting three phases in equilibrium. These correspond with bivariant, univariant, and invariant equilibria, respectively.

When the invariant *three-phase equilibria*, or as they are sometimes called reaction types, are examined, it is found that they are one of only two basic types, as shown in Figure 12.1. In the first type, (a), a high-temperature phase is no longer stable below the invariant temperature or, in reaction terms, the phase stable at high temperature dissociates into two other phases at the invariant temperature. In the other type, (b), a phase which is unstable at high temperature becomes stable below the invariant temperature. In reaction terms, two phases stable at high temperature react at the invariant temperature to form a new phase.

It is usual to go further and subdivide these two types of invariant equilibria into subcategories which depend on what kind of phases are involved at the invariant temperature. A complex, albeit hypothetical, phase diagram is shown in Figure 12.2. The names given to the different types of the subcategories of the two basic types are indicated on this diagram. The two other types of invariant point which have been met with previously, namely, a critical point and a congruent point, but which are not three-phase equilibria are also shown in this figure.

In this Chapter we will examine, more closely, the factors which determine the shape or topology of phase diagrams. There are constraints on what is and what is not acceptable and many published phase diagrams, determined by conventional means, have not complied with these constraints. With the increasing awareness of thermodynamic constraints, this is a less frequent occurrence nowadays.

A complete understanding of the observed topology of a given phase diagram requires consideration of:

1. The phase rule, which tells us the number of phases which can be present at *each* topographical feature.
2. A combinatorial analysis, which will tells us the *maximum possible number* of topographical features present in a given system.
3. Schreinemaker's rules, which tell us what are some of the topographical *possibilities and impossibilities*.

Materials Thermodynamics. By Y. Austin Chang and W. Alan Oates
Copyright 2010 John Wiley & Sons, Inc.

4. The Gibbs–Konovalov equation, which reveals the factors responsible for
   determining the slopes of phase boundaries

The application of all these concepts together is particularly useful for gaining
an understanding of multicomponent phase diagrams, but we can illustrate the
principles by confining ourselves to a consideration of binary systems only.

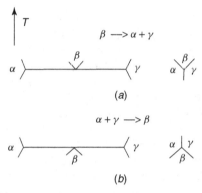

Figure 12.1   Two types of invariant equilibria in both $T-x_i$ and $T-\mu_i$ planes.

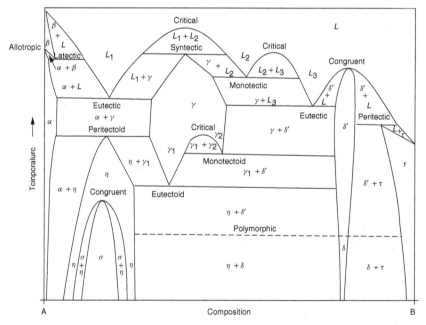

Figure 12.2   Hypothetical phase diagram which illustrates different types of invariant
equilibria. *Source: ASM Handbook Vol. 3*, 10th Edition.

12.1   GIBBS'S PHASE RULE

A principal use of the phase rule is to tell us how many phases are present at each topographical feature, each of which corresponds to a different value of F, the number of degrees of freedom, where F is the number of intensive variables which may be independently varied without changing the number of phases present at a selected point on the phase diagram.

At constant total pressure, where all species are mobile and in the absence of external fields and chemical reactions, $F = C - \phi + 1$. In a binary system, under these conditions, $F = 3 - \phi$. It follows that we may have only bivariant, univariant, and invariant equilibria corresponding to the presence of single-phase, two-phase, and three-phase regions on a phase diagram.

Two examples of topographical features which are obvious violations of the phase rule are shown in Figure 12.3. The first, illustrating a four-phase equilibrium, leads to $F = -1$ and the second suggests a three-phase equilibrium ($F = 0$) with the coexisting phases at different temperatures.

12.2   COMBINATORIAL ANALYSIS

The phase rule does not give any indication of how many of the different types of topographical features are present in a given system. A combinatorial analysis, on the other hand, permits the calculation of the upper limit to the number of different topographical features. Not all of the features included in this maximum total number are necessarily present in the phase diagram for a particular system.

In a system with C components and N total number of phases, the upper limit of invariant features, that is, where $F = 0$ and $\phi = C + 1$, may be calculated from the combinatorial rule, $\binom{N}{C+1}$. Similar evaluations can be made for univariant and bivariant features. In the case of phases which show a miscibility gap, it is necessary to include both segregated phases in the counting procedure. Also, the combinatorial calculation for the maximum possible number of the topographical features does not take the single-phase invariant congruent points and critical points into consideration.

Let us use the combinatorial formula for evaluating the maximum possible number of each feature in a binary system in which there are four different phases in total. From the phase rule, the maximum number of phases in equilibrium at

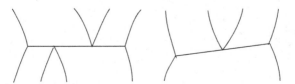

Figure 12.3   Phase rule violations in $T-x_i$ phase diagrams. *Source:* H. Okamoto and T. B. Massalski, *J. Phase Equilib.* **12** 148–168, 1991.

constant p (where $F = 0$) is 3. Therefore, the maximum possible total number of invariant features is $\binom{4}{3} = 4$, the maximum possible total number of univariant features is $\binom{4}{2} = 6$, and the maximum possible total number of bivariant features is $\binom{4}{1} = 4$. In more detail, for a system where the four phases are designated $\alpha, \beta, \gamma, \delta$, we can summarize all the possible topographical features as follows:

F	ϕ	Total	Combinations
0	3	4	$\alpha\beta\gamma, \alpha\beta\delta, \alpha\gamma\delta, \beta\gamma\delta$
1	2	6	$\alpha\beta, \alpha\gamma, \alpha\delta, \beta\gamma, \beta\delta, \gamma\delta$
2	1	4	$\alpha, \beta, \gamma, \delta$

It is obvious from this table that, for a given invariant feature, all possible univariant features may be calculated by combination of any two of the phases involved in the invariant equilibrium, for example, $\alpha\beta\gamma \rightarrow \alpha\beta, \alpha\gamma, \beta\gamma$ (see the sketches on the right-hand side of Figures 12.1a and 12.1b for a three-phase invariant point on a $T - \mu_i$ diagram).

The combinatorial calculation does not give information on which of the total number of possible features are likely to be stable and which features are likely to be metastable and therefore absent. The concept of the missing invariant points can be appreciated by considering a binary system in which there is a single stoichiometric compound and two eutectics on either side of the compound (see Fig. 12.4a).

Since there are four phases, the combinatorial result is that there is a total of four invariant equilibria (at constant p). But the actual phase diagram shows only two (plus the congruent point). In this case there must be two other metastable invariant points. The location of these metastable invariant equilibria are shown schematically on the $T - x_i$ and $T - \mu_i$ phase diagrams in Figure 12.4.

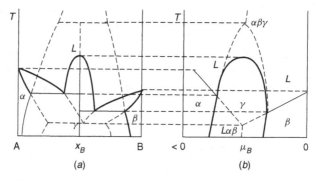

(a)   (b)

Figure 12.4 The $T - x_i$ and $T - \mu_i$ phase diagrams for system with stoichiometric compound. Locations of the metastable invariant equilibria are indicated on both diagrams.

12.3   SCHREINEMAKER'S RULES

Schreinemaker's rules are based on the concept that an equilibrium phase or phase assemblage must only have its stability field constricted by consideration of another yet more stable assemblage.

The rules are clearer in a field–field phase diagram than in a field–density phase diagram. In the former type:

1. Adjacent univariant curves around an invariant point are constrained to be <180°.
2. The metastable extension of a stable univariant curve to the opposite side of an invariant point must be into a different single-phase region.

Acceptable and unacceptable topologies are shown in Figure 12.5. It can be seen that application of the rules to extrapolations is clearer in the field–field phase diagram than they are in the field–density diagram. The correctness of their application to $T - x_B$ phase diagrams can be appreciated from $G - x_B$ curves, such as those shown in Figure 12.6. It can be seen there that the phase boundaries for the metastable α/β equilibrium lie inside those for the stable liq/α and liq/β equilibria.

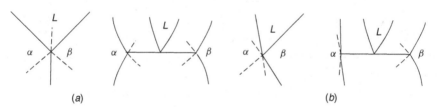

Figure 12.5   Stable phase boundaries and their extrapolations into metastable regions in both $T - \mu_i$ and $T - x_i$ phase diagrams: (*a*) acceptable; (*b*) unacceptable.

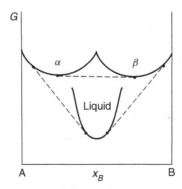

Figure 12.6   Application of Schreinemaker's rules can be appreciated from $G - x_B$ curves.

12.4   THE GIBBS–KONOVALOV EQUATIONS

The phase rule, combinatorics, or the application of Schreinemaker's rules will not tell us if there actually is a miscibility gap, a congruently melting compound, or anything about the actual juxtaposition of the different phases and location of the invariant equilibria in a particular system. Only knowledge of the thermodynamic properties of the phases involved can do this. The thermodynamic properties also determine the slopes of the phase boundaries in both $T-x_i$ and $T-\mu_i$ phase diagrams and it is on this aspect that we will concentrate here.

The phase diagrams shown in Figures 12.7 and 12.8 are examples which exhibit some rather unusual features.

The diagram in Figure 12.7a shows an increase in solid solubility with falling temperature which is then followed by a decrease in the solubility. The solute whose solubility increases with decreasing temperature is said to have *retrograde* solubility. The phase diagram shown in Figure 12.7b is unusual in that it

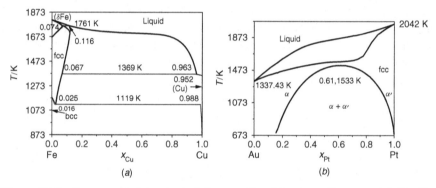

Figure 12.7   Some unusual topological features found in phase diagrams: (a) Cu–Fe; (b) Au–Pt.

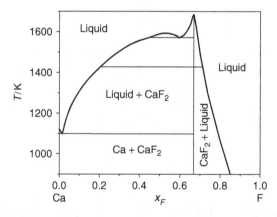

Figure 12.8   Apparent sharp congruent point in Ca–F system.

shows an almost completely flat solidus. That in Figure 12.8 shows a congruent point which is *apparently* quite sharp. These and other topological features can be understood from a thermodynamic analysis of the factors which control the slopes of phase boundaries. Such an analysis can prevent an experimentalist from interpolating/extrapolating his or her results in an erroneous manner in the construction of the final phase diagram. This kind of analysis is unimportant in calculated phase diagrams since thermodynamic consistency is automatic.

The object is to determine the factors which are responsible for the slopes of phase boundaries on isobaric phase diagrams of both the $T - \mu_i$ and $T - x_i$ type, that is, we wish to calculate

$$\left(\frac{\partial T}{\partial \mu_i} \right)_p \quad \text{and} \quad \left(\frac{\partial T}{\partial x_i^\alpha} \right)_p$$

Note that, in the second case, it is necessary to specify both the component and the phase since there are two phases of different composition, that is, two phase boundaries, to consider.

Both of the above partial derivatives refer to univariant equilibria and, along the univariant phase boundaries, the component chemical potentials are equal in both phases. The total differentials of μ_i for each component in each phase are also equal:

$$d\mu_A^\alpha = d\mu_A^\beta = d\mu_A \tag{12.1}$$

$$d\mu_B^\alpha = d\mu_B^\beta = d\mu_B \tag{12.2}$$

We will use these relations in the derivations given below.

12.4.1   Slopes of $T - \mu_i$ Phase Boundaries

The Gibbs–Duhem equations for both phases along the phase boundary are

$$x_A^\alpha \, d\mu_A + x_B^\alpha \, d\mu_B + S^\alpha \, dT + V^\alpha \, dp = 0 \tag{12.3}$$

$$x_A^\beta \, d\mu_A + x_B^\beta \, d\mu_B + S^\beta \, dT + V^\beta \, dp = 0 \tag{12.4}$$

At constant p, we can eliminate $d\mu_A$ from these equations by multiplying with the appropriate factors and then subtracting:

$$x_A^\alpha x_A^\beta \, d\mu_A + x_B^\alpha x_A^\beta \, d\mu_B + x_A^\beta S^\alpha \, dT = 0 \tag{12.5}$$

$$x_A^\alpha x_A^\beta \, d\mu_A + x_A^\alpha x_B^\beta \, d\mu_B + x_A^\alpha S^\beta \, dT = 0 \tag{12.6}$$

$$(x_A^\alpha x_B^\beta - x_B^\alpha x_A^\beta) \, d\mu_B + (x_A^\alpha S^\beta - x_A^\beta S^\alpha) \, dT = 0 \tag{12.7}$$

and, since $x_A^\alpha x_B^\beta - x_B^\alpha x_A^\beta = x_A^\alpha - x_A^\beta$, we have

$$\left(\frac{\partial T}{\partial \mu_B}\right)_p = \frac{x_A^\alpha - x_A^\beta}{x_A^\beta S^\alpha - x_A^\alpha S^\beta} \tag{12.8}$$

There is, of course, an analogous equation for $dT/d\mu_A$.

Example 12.1 Phase Boundary Slopes

Confirm (12.8) for the following system at 600 K. The calculated phase diagram is shown in Figure 12.9. This phase diagram was obtained by taking the liquid phase to be strictly regular and the solid phase to be strictly subregular. The pure liquids are chosen as reference states:

$$\mu_A^{\circ,l} = \mu_B^{\circ,l} = 0$$

$$\mu_A^{\circ,s} - \mu_A^{\circ,l} = -9000 + 10T \text{ J mol}^{-1}$$

$$\mu_B^{\circ,s} - \mu_B^{\circ,l} = -12{,}000 + 10T \text{ J mol}^{-1}$$

$$G^E(l) = 5000 x_A x_B \text{ J mol}^{-1}$$

$$G^E(s) = x_A x_B (10{,}000 + 10{,}000(x_A - x_B)) \text{ J mol}^{-1}$$

Using the phase boundary values given in Figure 12.9 for $T = 600$ K, we have

$$\frac{dT}{d\mu_A} = \frac{0.9866 - 0.3033}{0.3033(-9.409) - 0.9866(-4.898)} = 0.345$$

$$\frac{dT}{d\mu_B} = \frac{0.0134 - 0.697}{0.6967(-9.409) - 0.0134(-4.898)} = 0.105$$

which agree well with the slopes given in Figure 12.10a.

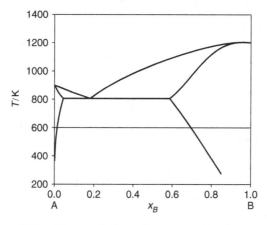

Figure 12.9 Calculated phase diagram used in the example.

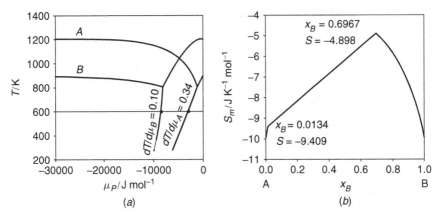

Figure 12.10 (a) $d\mu/dT$ and (b) $S_m(600 \text{ K}) - x_B$ diagrams for system shown in Figure 12.9.

12.4.2 Slopes of $T - x_i$ Phase Boundaries

The derivation of the equations for the phase boundary slopes is again based on using procedures very similar to those used in the derivation of the Clausius–Clapeyron. Along the equilibrium phase boundary between the two phases we have

$$\mu_i^\alpha - \mu_i^\beta = 0 \qquad d\mu_i^\alpha - d\mu_i^\beta = 0$$

At constant total p, the total differential for μ_i is

$$d\mu_i = \left(\frac{\partial \mu_i}{\partial x_i}\right)_T dx_i + \left(\frac{\partial \mu_i}{\partial T}\right)_{x_i} dT \qquad (12.9)$$

The two partial derivatives may now be obtained:

(a) From the tangent intercept relation at constant p:

$$\mu_i = G_m + (1 - x_i)\left(\frac{\partial G_m}{\partial x_i}\right)$$

so that

$$\left(\frac{\partial \mu_i}{\partial x_i}\right)_T = \left(\frac{\partial G_m}{\partial x_i}\right)_T + (1 - x_i)G_{x_i x_i}(T) - \left(\frac{\partial G_m}{\partial x_i}\right)_T$$
$$= (1 - x_i)G_{x_i x_i}(T)$$

(b)
$$\left(\frac{\partial \mu_i}{\partial T}\right)_{x_i} = -S_i(x_i)$$

where we have used $G_{x_i x_i}(T)$ for the second derivative of G_m with respect to x_i. Note that, since $\mu_B - \mu_A = dG_m/dx_B = -dG_m/dx_A$ and in the case of the second derivatives,

$$G_{x_A^\alpha x_A^\alpha}(T) = G_{x_B^\alpha x_B^\alpha}(T) = G_{xx}^\alpha(T)$$

We now have, for the two equations for $d\mu_i^\alpha - d\mu_i^\beta = 0$,

$$x_B^\alpha G_{xx}(T)\,dx_A^\alpha - S_A^\alpha(x_A^\alpha)\,dT - x_B^\beta G_{xx}(T)\,dx_A^\beta + S_A^\beta(x_A^\beta)\,dT = 0$$

$$x_A^\alpha G_{xx}(T)\,dx_B^\alpha - S_B^\alpha(x_B^\alpha)\,dT - x_A^\beta G_{xx}(T)\,dx_B^\beta + S_B^\beta(x_B^\beta)\,dT = 0$$

Remembering that $dx_A = -dx_B$ in either phase, we now have two equations from which dT/dx_B^α and dT/dx_B^β can be obtained. These are the *Gibbs–Konovalov* equations for the slopes of the two phase boundaries for a two-phase equilibrium *at constant p, T*:

$$\frac{dT}{dx_B^\alpha} = \frac{(x_B^\alpha - x_B^\beta) G_{xx}^\alpha(T)}{x_B^\beta \Delta_\alpha^\beta S_B + x_A^\beta \Delta_\alpha^\beta S_A}$$

$$\frac{dT}{dx_B^\beta} = \frac{(x_B^\alpha - x_B^\beta) G_{xx}^\beta(T)}{x_B^\alpha \Delta_\alpha^\beta S_B + x_A^\alpha \Delta_\alpha^\beta S_A}$$

Since, along the phase boundary, $\mu_i^\alpha = \mu_i^\beta$, it follows that $\Delta_\alpha^\beta H_i = T \Delta_\alpha^\beta S_i$, so that this substitution may be made in the above equations. This is the way that the equations are most often presented for the isobaric conditions being considered:

$$\boxed{\frac{dT}{dx_B^\alpha} = \frac{(x_B^\alpha - x_B^\beta) T G_{xx}^\alpha(T)}{x_B^\beta \Delta_\alpha^\beta H_B + x_A^\beta \Delta_\alpha^\beta H_A}} \qquad (12.10)$$

$$\boxed{\frac{dT}{dx_B^\beta} = \frac{(x_B^\alpha - x_B^\beta) T G_{xx}^\beta(T)}{x_B^\alpha \Delta_\alpha^\beta H_B + x_A^\alpha \Delta_\alpha^\beta H_A}} \qquad (12.11)$$

Example 12.2 Phase Boundary Slopes
Confirm (12.10) and (12.11) for the same system as used in the previous example at 600 K. The approximate slopes of the boundaries on the $T-x_B$ phase diagram are indicated on Figure 12.11a. The values of $G_{xx}^\alpha(T)$ [$= G_{xx}^\beta(T)$ in this case] are given in Figure 12.11b.

The slopes given in the figure are only approximate:

$$\frac{dT}{dx_B^\alpha} = \frac{(0.0134 - 0.6967)2.989 \times 10^5}{0.6967(-32.85) - 0.3033(-9.807)} = +10{,}252$$

$$\frac{dT}{dx_B^\beta} = \frac{(0.0134 - 0.6967)0.272 \times 10^5}{0.0134(-32.85) - 0.9866(-9.807)} = -2011$$

12.4.3 Some Applications of Gibbs–Konovalov Equations

We may now consider the application of (12.8) and the Gibbs–Konovalov equations, (12.10) and (12.11), in order to explain the slopes of phase boundaries on both $T-\mu_i$ and $T-x_i$ phase diagrams. We concentrate on the features illustrated in the hypothetical phase diagram shown in Figure 12.12.

We use, as is apparent from Figure 12.11b, that, in the infinitely dilute solutions, $G_{xx} \to \infty$. At other compositions, it may be either positive or negative. We have previously noted that this second derivative is zero at a critical point.

1. *Congruent and Critical Points*   At both congruent and critical points (the points indicated by 1 in Fig. 12.12), $x_B^\alpha = x_B^\beta$, so it follows, from (12.8), (12.10), and (12.11), that

$$\frac{dT}{d\mu_B} = \frac{dT}{dx_B^\alpha} = \frac{dT}{dx_B^\beta} = 0$$

At a liquid/solid intermediate-phase transformation, both the liquidus and solidus have zero slope at the actual congruent point. Although it may appear that the congruent point in Figure 12.8 is quite sharp, it must actually be rounded. This can be demonstrated through the use of appropriate modeling equations for the

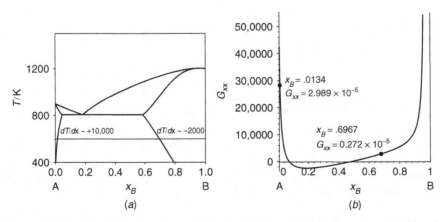

Figure 12.11   Phase boundary slopes and values of G_{xx} at 600 K for system considered: (*a*) slopes of $T-x_i$ phase diagram; (*b*) G_{xx} at phase boundaries.

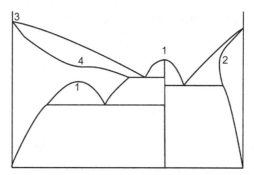

Figure 12.12 Some topological features which may be understood from Gibbs–Konovalov equations.

liquid phase. A sharp congruent point like that shown requires a deep minimum in the $\Delta_{mix}G_m - x_i$ curve. The zero-slope requirement at a congruent point also applies to critical points. At the pure component ends of the phase diagram, where x_B^α is also equal to x_B^β, the requirement that $dT/d\mu_B = dT/dx_B^\alpha = dT/dx_B^\beta = 0$ no longer holds since, at these end compositions, G_{xx} is no longer finite.

2. *Azeotropy*   A further consequence of (12.8), (12.10), and (12.11) is to exclude phase diagrams like the one in the accompanying sketch. The liquidus and solidus can only touch ($x_B^\alpha = x_B^\beta$) when the slope is zero. They can, therefore, only touch at a maximum or a minimum. Such points are called *azeotropic points*, the best known example being that for the liquid/vapor equilibrium for water and alcohol.

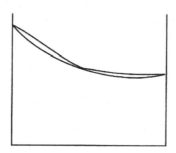

3. *Retrograde Solubility*   An example of retrograde solubility was given in Figure 12.7a and at the point marked as 2 in Figure 12.12. At the retrograde point, the solidus phase boundary is vertical, that is, $dT/dx_B^\alpha = \infty$. This can only occur if the denominator is zero, that is, when

$$x_B^\alpha \, \Delta_I^\beta S_B + x_A^\alpha \, \Delta_I^\beta S_A = 0 \qquad (12.12)$$

4. *Inflection Points*   At the critical point $G^{\alpha}_{xx} = 0$ so that, in the neighborhood of this critical point, $G^{\alpha}_{xx} \approx 0$, and since, from (12.10) and (12.11)

$$\frac{dT}{dx_B(s)} \quad \propto \quad G_{xx}(s)$$

it follows that if the solidus is fairly close to the critical point for the solid-state miscibility gap as in the case of the points marked 1 and 4 in Figure 12.12 and in Figure 12.7b for the (Au,Pt) system, then

$$\frac{dT}{dx^{\alpha}_B} \approx 0$$

that is, under these conditions, an inflection point will occur in the solidus (the solidus can never be horizontal except at an azeotropic point). An accompanying inflection point in the liquidus is not necessary—this would require $G_{xx}(l) \approx 0$, which would only occur if the liquid phase was itself close to giving rise to phase separation.

5. *Dilute Solutions*   For $x_B \to 0, x_A \to 1$, we can take the ideal entropy so that $G^{\alpha}_{xx} \to RT/x^{\alpha}_B$. Under these conditions, the Gibbs–Konovalov equations simplify to

$$\frac{dT}{dx^{\alpha}_B} = \frac{(x^{\alpha}_B - x^{\beta}_B)RT}{x^{\alpha}_B \, \Delta^{\beta}_{\alpha} S_A} \tag{12.13}$$

$$\frac{dT}{dx^{\beta}_B} = \frac{(x^{\alpha}_B - x^{\beta}_B)RT}{x^{\beta}_B \, \Delta^{\beta}_{\alpha} S_A} \tag{12.14}$$

If we invert these and subtract,

$$\frac{dx^{\alpha}_B}{dT} - \frac{dx^{\beta}_B}{dT} = \frac{\Delta^{\beta}_{\alpha} S_A}{RT} \tag{12.15}$$

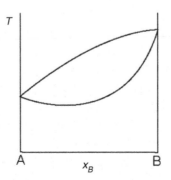

If the α phase is the low-temperature phase, then $\Delta_I^\beta S_A$ will be positive, which means that the difference in the reciprocal slopes of the coexistence in very dilute solution, (12.15), must be positive. If both are positive, say, $dT/dx_B^\alpha = 2$ and $dT/dx_B^\beta = 4$, then the difference of the reciprocals is $\frac{1}{2} - \frac{1}{4} = \frac{1}{4}$. If both are negative, say, $dT/dx_B^\alpha = -4$ and $dT/dx_B^\beta = -2$, then the difference of the reciprocals is $-\frac{1}{4} + \frac{1}{2} = \frac{1}{4}$. But if one is positive and the other negative, say $dT/dx_B^\alpha = -2$ and $dT/dx_B^\beta = 4$, then the difference of the reciprocals is $-\frac{1}{2} - \frac{1}{4} = -\frac{3}{4}$, which is negative. This means that a phase diagram like the one in the sketch on page 161, where the liquidus on the A-rich end has a positive slope while the solidus has a negative slope, is impossible.

Summary   The computer calculation of phase diagrams ensures that all thermo-dynamic requirements are met, that is, there are no errors in calculated phase diagrams due to any failure in meeting the requirements of the phase rule, Schreine-maker's rules, or the Gibbs–Konovalov equation. Nevertheless, many published phase diagrams have not been computed but are based on experimental results. They are thus prone to contain mistakes which are thermodynamically incorrect.

EXERCISES

12.1   The hypothetical phase diagram in Figure 12.13 contains many mistakes which are indicated by a letter. For each of these mistakes, list the type of error in terms of its failure to be in line with the requirements of:

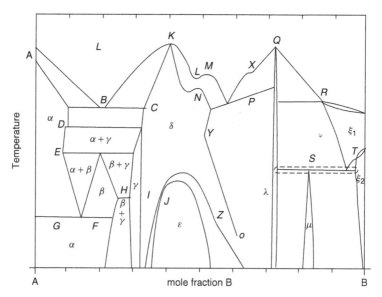

Figure 12.13   Hypothetical phase diagram containing many topological errors.

  (i) The phase rule

 (ii) Schreinemakers's rules

(iii) The Gibbs–Konovalov's equation

12.2 Using the data below, calculate the phase diagram, evaluate the retrograde point, and confirm (12.12).

  Be sure to use the same reference state for both phases when evaluating the entropy difference terms in (12.12).

$$\mu_A^{\circ,1} = \mu_B^{\circ,1} = 0$$

$$\mu_A^{\circ,s} - \mu_A^{\circ,1} = -22{,}500 + 15T \text{ J mol}^{-1}$$

$$\mu_B^{\circ,s} - \mu_B^{\circ,1} = -7500 + 15T \text{ J mol}^{-1}$$

$$G^E(1) = 10{,}000 x_A x_B \text{ J mol}^{-1}$$

$$G^E(s) = 15{,}000 x_A x_B \text{ J mol}^{-1}$$

13 Solution Phase Models I: Configurational Entropies

Up until this point we have discussed the representation of the thermodynamic properties of solution phases by using polynomials such as the Redlich–Kister equation:

$$G^{\mathrm{E}} = x_A x_B [L_0 + L_1(x_A - x_B) + L_2(x_A - x_B)^2 \ldots]$$

where the parameters L_i may be temperature dependent. Such representations do, however, have some limitations:

(a) Fitting sparse experimental results to polynomials, particularly high-order polynomials, can be dangerous, particularly when it is necessary to extrapolate the equations outside the region (of composition and/or temperature) where the fitting was carried out.

(b) The experimental results for some phases are unsuited to polynomial representation. This is particularly true of some intermediate phases, which are formed when there is a strong chemical affinity between the alloying elements involved. These phases progress from being fully ordered at low temperatures, as illustrated schematically in Figure 13.1b, to being disordered at high temperatures (as in Fig. 13.1a). As we will see later, $\Delta_{\mathrm{mix}} H_m$ changes quite markedly with composition and temperature due to the ordering process, and not in the simple way that can be represented reliably by a Redlich–Kister type of equation.

(c) To date, we have been concerned only with substitutional alloys, as sketched in Figures 13.1a and b. Sometimes, however, we encounter interstitial solid solution phases, as represented in Figure 13.1c. In the latter, the metal host atoms form one sublattice while the interstitial sites, formed within the metal atom sublattice, are occupied by the interstitial atoms. Since the metal atom cage must always be present to provide the interstitial sites, it is clear that this type of solution cannot exist across the whole composition range from one pure element to the other, as it does in the case of a substitutional solution. A (mathematical) modeling equation like the Redlich–Kister equation can be modified from that

Materials Thermodynamics. By Y. Austin Chang and W. Alan Oates
Copyright 2010 John Wiley & Sons, Inc.

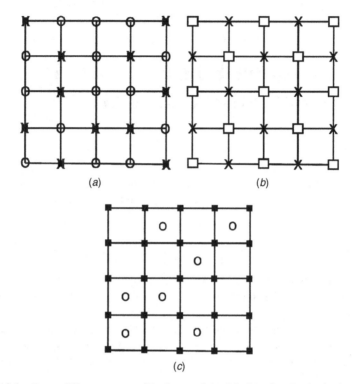

Figure 13.1   Some different types of lattice models: (*a*) disordered substitutional phase, all sites available for all types of atoms; (*b*) fully ordered intermediate phase, separate sublattices for each type of atom; (*c*) interstitial solution phase, interstitial atoms occupy their own sublattice.

used for substitutional solution phases as long as the interstitial solid solution is disordered. But interstitial solid solutions can also undergo ordering transitions and, again, a polynomial representation of their thermodynamic properties is unsatisfactory.

It is for these reasons that we need to supplement a wholly mathematical approach to modeling equations, by introducing a little more physics into the modeling process with the hope of obtaining a better representation of the experimental (or theoretically calculated) results. In particular, we wish to be able to handle sublattice phases of either the interstitial solution- or intermediate-phase type and we particularly wish to be able to have a model which will lead to a representation which is suitable for those phases which undergo order/disorder transitions.

The physicochemical factors which contribute to solid solution and compound formation between metals were summarized by Hume-Rothery in the 1930s, who found the following *atomic* properties to be important:

(i) *Atomic Size Factor*   If the atomic radii of the elements differ by more than $\approx 15\%$, solid solution formation is not favored.

(ii) *Crystal Structure Factor*   For appreciable solid solubility, the crystal structure of the component metals must be the same.

(iii) *Electrochemical Factor*   The greater the electronegativity difference between the components, the greater the tendency to form intermetallic compounds.

(iv) *Valence Factor*   For extensive solid solubility, the components should have the same valences. A metal will dissolve another metal of higher valency more than it will one of lower valency.

Hume-Rothery discovered these rules from the relatively small amount of information available to him some 80 years ago. Very large databases are now available for metal properties, and when these are linked with a machine learning system (data mining), predictions can be made about the structural and stability properties of alloys. It is interesting to note that the atomic properties used in correlating the information on alloy phase stability in one such large database are very similar to the ones used by Hume-Rothery: size, atomic (Mendeleev) number, valence electron, electrochemical, angular-valence orbital, and cohesive energy factors.

In obtaining such correlations, it is being assumed that alloy properties can be understood wholly in terms of the component *atomic* properties, which would seem to be a little optimistic. Although the elemental properties clearly give some clues as to the important factors which contribute to alloy thermodynamic mixing properties, they cannot lead to what is required in the present instance, namely, an analytical expression for these properties as a function of temperature and compositions. Recent first-principles calculations have been quite successful in calculating total energies for binary compounds at 0 K. Some improvements are still required, however, in the calculation of high-temperature Gibbs energies and in extending the calculations to disordered phases. As a result, a less ambitious phenomenological or empirical approach to solution-phase modeling is still necessary. Some physics is introduced but the model parameters are obtained, not by calculation, but by their optimization with respect to the available experimental or calculated results to give the best description of the experimental Gibbs energies. This empirical approach will probably be necessary for some years to come, particularly since it is multicomponent alloys which are of primary technological significance. First-principles calculations for these would seem some considerable way off.

In order to obtain the desired analytical representation, we simplify things by assuming that the Gibbs energy of mixing can be separated additively into:

(i) *Configuration independent* (abbreviation conf indep) contributions. These depend only on the alloy composition and not on how the component atoms are distributed in the alloy; that is, these contributions are identical for the fully ordered and for the random solution. A large part of any atomic size mismatch contributions and some part of electronic band contributions *may* fall into this category.

(ii) *Configuration dependent* (abbreviation conf) contributions. These do depend on how the component atoms are distributed in the alloy. An ordered alloy has different properties from a disordered alloy from these contributions. In other words, these are *local environment effects*. Apart from the mixing properties themselves, we might expect that at least part of the vibrational properties of an atom will be sensitive to the local environment.

With this assumption of the two contributions being independent of one another, we can write the solution-phase Gibbs energy as follows:

$$G_m = G_m^{conf} + G_m^{confindep} \tag{13.1}$$

It should be emphasized that this factorization of G_m can be an oversimplification in that the the two contributions may be *coupled*.

Now $G_m^{conf} = H_m^{conf} - T S_m^{conf}$ and here we will make another big assumption, namely, that these two contributions to G_m^{conf} can be treated independently, so that we may now write

$$G_m = \left[H_m^{conf} - T S_m^{conf} \right] + G_m^{confindep}$$

In the remainder of this chapter we concentrate on the calculation of S_m^{conf} for the three types of solution phase shown in Figure 13.1.

13.1 SUBSTITUTIONAL SOLUTIONS

We have previously met the Boltzmann equation, applicable to systems of fixed energy:

$$S = k_B \log_e W \tag{13.2}$$

This equation indicates that we can evaluate the entropy from a calculation of the number of microstates available to a system. For any solid substance the available states will include nuclear, electronic (including nuclear and electronic spin), and vibration energy levels—*thermal* disorder. For the moment, however, we are concerned only with *configurational* disorder—the more numerous the possible ways of arranging the atoms on a lattice, the higher the configurational entropy.

In a substitutional solid solution, all the lattice sites are considered to be available for occupation by every atom. If the energy of mixing is taken to have the constant value of zero, it is then straightforward to calculate the number of microstates for a *random* binary solution. This random approximation is referred to as the *point, zeroth, or regular solution* approximation. When N_A and N_B particles are distributed randomly on $N_s = N_A + N_B$ sites, then

$$W = \frac{N_s!}{N_A! \, N_B!} \tag{13.3}$$

Use Stirling's approximation for the factorial of a large number,

$$\log_e x! \approx x \log_e x - x$$

we obtain from (13.2) and (13.3)

$$\frac{S^{\text{conf}}}{k_B} = N_s \log_e N_s - N_A \log_e N_A - N_B \log_e N_B \qquad (13.4)$$

Since $W = 1$ for the pure components, the entropy for the random solution is the same as integral mixing entropy *per mole of sites* for the alloy, $\Delta_{\text{mix}} S_m^{\text{conf}}$:

$$-\frac{\Delta_{\text{mix}} S_m^{\text{conf}}}{N_s k_B} = x_A \log_e x_A + x_B \log_e x_B \qquad \text{random subst. soln.} \qquad (13.5)$$

The partial molar entropies can be obtained by differentiating (13.4):

$$\frac{S_A^{\text{conf}}}{k_B} = \left(\frac{\partial S/k_B}{\partial N_A} \right)_{N_B} = \log_e N_s - \log_e N_A = -\log_e x_A$$

Using i to represent any component, the partial molar entropy of mixing is given by

$$-\frac{\Delta_{\text{mix}} S_i^{\text{conf}}}{R} = \log_e x_i \qquad \text{random subst. soln.} \qquad (13.6)$$

As we have seen previously, such a random substitutional alloy with a zero energy of mixing will obey Raoult's law.

13.2   INTERMEDIATE PHASES

Consider the simple case where the intermediate phase can be described as being comprised of only two sublattices, similar to that shown schematically in Figure 13.1b. The usual notation for representing such an intermediate phase is

$$(A,B)_{f^\alpha} : (B,A)_{f^\beta}$$

where A is the predominant species on the first sublattice and B on the second sublattice.

We will use i to refer to any sublattice constituent and (j) to refer to any sublattice. Thus $f^{(j)}$, which is the fraction of the total sites which are on sublattice (j), is given by

$$f^{(j)} = \frac{N_s^{(j)}}{N_s} \qquad \text{where} \qquad \sum_{(j)} f^{(j)} = 1$$

with $N_s^{(j)}$ being the number of sites on sublattice (j) and N_s the total number of sites.

It is also useful to define the *sublattice mole fraction* as

$$y_i^{(j)} = \frac{N_i^{(j)}}{N_s^{(j)}}$$

The relation between the sublattice mole fractions and the mole fraction can be obtained as follows:

$$x_i = \frac{N_i}{N_s} = \frac{N_i^\alpha + N_i^\beta}{N_s} = \frac{f^\alpha N_i^\alpha}{N_s^\alpha} + \frac{f^\beta N_i^\beta}{N_s^\beta}$$

or, in general,

$$\boxed{x_i = \sum_{(j)} \sum_i f^{(j)} y_i^{(j)}} \tag{13.7}$$

Let us now obtain the expression for the mixing entropy for this case. Analogous to the disordered alloy, we will assume that there is *random* mixing on each individual sublattice. This is known as the Bragg–Williams (BW) approximation. Under these circumstances, the energy of mixing is fixed at constant composition, so that the Boltzmann equation can be applied.

The equation for the thermodynamic probability is now given by

$$W = \left[\frac{N_s^\alpha!}{N_A^\alpha! \, N_B^\alpha!} \times \frac{N_s^\beta!}{N_A^\beta! \, N_B^\beta!} \right] \tag{13.8}$$

From the Boltzmann equation and using the Stirling approximation and generalizing,

$$\frac{S^{conf}}{k_B} = \sum_{(j)} \left[N_s^{(j)} \log_e N_s^{(j)} - \sum_i N_i^{(j)} \log_e N_i^{(j)} \right] \tag{13.9}$$

$$= \sum_{(j)} \sum_i N_i^{(j)} \log_e y_i^{(j)} \tag{13.10}$$

Dividing by N_s to obtain the entropy per lattice site,

$$-\frac{S^{conf}}{k_B N_s} = \sum_{(j)} f^{(j)} \sum_i y_i^{(j)} \log_e y_i^{(j)} \tag{13.11}$$

Since $W = 1$ for the pure elements, this is also the molar mixing entropy:

$$-\frac{\Delta_{mix} S_m^{conf}}{R} = \sum_{(j)} f^{(j)} \sum_i y_i^{(j)} \log_e y_i^{(j)} \qquad \text{BW approx.} \qquad (13.12)$$

The partial molar entropies are given by

$$-\frac{\Delta_{mix} S_i^{conf}}{R} = \sum_{(j)} f^{(j)} \sum_i \log_e y_i^{(j)} \qquad \text{BW approx.} \qquad (13.13)$$

These expressions have the correct high-temperature limit of giving the same answer as for the random substitutional solid solution. When there is only one sublattice, $f^{(j)} = 1$, and then, from (13.7), $y_i^{(j)} = x_i$, and we recover (13.5) from (13.9).

We can also readily evaluate (13.12) and (13.13) for the BW most ordered state.

At the stoichiometric composition, AB, and in the fully ordered condition, the first sublattice is occupied by A atoms only and the second sublattice by B atoms only, that is, $y_B^\alpha = 0$, $y_B^\beta = 1$. On moving to the B-rich side, we can only increase y_B^α since the second sublattice is already filled. The variation of y_B^α and y_B^β as a function of x_B, over the whole composition range, is shown in Figure 13.2. It can be seen that, for this most ordered state,

$$y_B^\alpha = 0 \qquad y_B^\beta = 2x_B \qquad \text{for } x_B \le \tfrac{1}{2}$$
$$y_B^\alpha = 2x_B - 1 \qquad y_B^\beta = 1 \qquad \text{for } x_B > \tfrac{1}{2}$$

Similar equations exist for y_A and we can then substitute these values for the sublattice mole fractions into (13.12) and (13.13) in order to calculate the integral and partial molar mixing entropies for this most ordered state. The following relations are obtained for the integral and partial quantities (for component B):

$$-\frac{\Delta_{mix} S_m^{conf}}{R} = \begin{cases} \tfrac{1}{2}[2x_B \log_e(2x_B) + (1 - 2x_B)\log_e(1 - 2x_B)] & x_B < \tfrac{1}{2} \quad (13.14) \\ \tfrac{1}{2}[2x_A \log_e(2x_A) + (1 - 2x_A)\log_e(1 - 2x_A)] & x_B > \tfrac{1}{2} \quad (13.15) \end{cases}$$

$$-\frac{\Delta_{mix} S_B}{R} = \begin{cases} \log_e 2x_B - \tfrac{1}{2}\log_e(1 - 2x_B) & x_B < \tfrac{1}{2} \quad (13.16) \\ \tfrac{1}{2}\log_e(2x_B - 1) & x_B > \tfrac{1}{2} \quad (13.17) \end{cases}$$

The results for the integral and partial quantities for both the random and most ordered states are shown in Figures 13.3a and 13.3b, respectively. These represent the low- and high-temperature limits for these quantities.

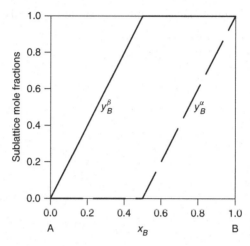

Figure 13.2 Variation of sublattice mole fraction with composition for BW most ordered state in two-sublattice intermediate phase.

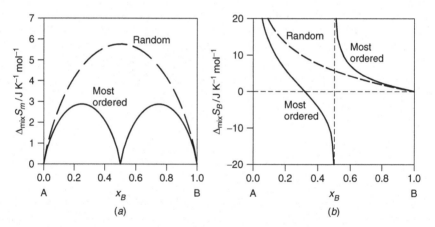

Figure 13.3 Integral and partial molar configurational mixing entropies for two-sublattice intermediate phase in BW most ordered state: (a) integral $\Delta_{mix}S_m$; (b) partial $\Delta_{mix}S_B$.

13.3 INTERSTITIAL SOLUTIONS

Now consider a binary interstitial solid solution with two sublattices: one, α, occupied by the metal atoms, M, and another, β, formed by the spaces lying between the metal atoms and in which an element, I, occupies some fraction of these interstitial sites. This model can be represented by the following notation:

$$(M)_{f^\alpha} : (I, Va)_{f^\beta}$$

where Va represents a vacancy on the interstitial sublattice.

We may go straight to using (13.9) and consider the *random* mixing of atoms and vacancies on the interstitial sublattice:

$$\frac{S^{\text{conf}}}{k_B} = N_s^\beta \log_e N_s^\beta - [N_I^\beta \log_e N_I^\beta + N_{\text{Va}}^\beta \log_e N_{\text{Va}}^\beta] \qquad (13.18)$$

We may eliminate N_{Va} and express S in terms of only the chemical components by using $N_{\text{Va}}^\beta = N_s^\beta - N_I^\beta$:

$$\frac{S^{\text{conf}}}{k_B} = N_s^\beta \log_e N_s^\beta - \left[N_I^\beta \log_e N_I^\beta + (N_s^\beta - N_I^\beta) \log_e (N_s^\beta - N_I^\beta) \right] \qquad (13.19)$$

and can express this, per mole of interstitial sites, by dividing by N_s^β:

$$-\frac{S^{\text{conf}}}{k_B N_s^\beta} = \left[y_I^\beta \log_e y_I^\beta + (1 - y_I^\beta) \log_e (1 - y_I^\beta) \right] \qquad (13.20)$$

which has the same analytical form as the equation for a random substitutional solid solution.

Equation (13.9) can be differentiated to obtain the partial quantity for the interstitial element I. Per atom,

$$\frac{S_I^{\text{conf}}}{k_B} = \left(\frac{\partial S^{\text{conf}}/k_B}{\partial N_I} \right)_{N_M} = -\log_e N_I^\beta + \log_e (N_s^\beta - N_I^\beta)$$

$$= -\log_e \left(\frac{y_I^\beta}{1 - y_I^\beta} \right) \qquad (13.21)$$

and for the partial molar quantity,

$$\frac{S_I^{\text{conf}}}{R} = -\log_e \left(\frac{y_I^\beta}{1 - y_I^\beta} \right) \qquad (13.22)$$

In order to obtain the partial quantity for M we must express N_s^β in terms of N_M. Let $N_s^\beta = \zeta N_M$, where ζ is the number of interstitial sites per metal atom [this means that $f^\alpha = 1/(1 + \zeta)$ and $f^\beta = \zeta/(1 + \zeta)$]. Using this in (13.18) and differentiating, per atom,

$$\frac{S_M^{\text{conf}}}{k_B} = \left(\frac{\partial S^{\text{conf}}/k_B}{\partial N_M} \right)_{N_I} = \log_e \zeta N_M - \log_e (\zeta N_M - N_I^\beta)$$

$$= -\zeta \log_e (1 - y_I^\beta) \qquad (13.23)$$

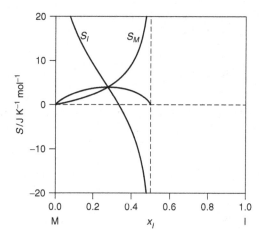

Figure 13.4 Partial molar entropies of mixing for random interstitial solid solution with $\zeta = 1$.

and for the partial molar quantity,

$$\frac{S_M^{\text{conf}}}{R} = -\zeta \, \log_e(1 - y_I^\beta) \tag{13.24}$$

Note that, even though the metal sublattice remains totally unaffected by the mixing process, which occurs only on the interstitial sublattice, there is a change in its partial molar property with composition.

The partial molar entropies for this random interstitial solid solution are shown in Figure 13.4a for the case where $\zeta = 1$ (which corresponds with the octahedral interstitial sites in the fcc lattice). The partial molar quantity for the interstitial component goes to plus and minus infinity, which is quite different behavior from that for the substitutional alloy. It is clear that a random interstitial solution will not be expected to obey Raoult's law; that is, obeying Raoult's law is *not* an appropriate definition of an ideal solution for an interstitial solid solution.

A random interstitial solid solution whose partial molar configurational entropies are given by the above equations is said to form a *Langmuir* ideal solution.

We should note again that a thermodynamicist may define an ideal solution in any way that he or she wants. We have now met two different types and definitions: Raoultian and Langmuir ideal solutions. The major point is that each is defined to be a limit to which real solutions of that type may approach. A Raoultian ideal solution is appropriate for substitutional phases and a Langmuir ideal solution for interstitial phases.

EXERCISES

13.1 Starting from the Boltzmann equation, (13.2), derive $\Delta_{\text{mix}} S_m^{\text{conf}}$ for a random solution mixture of $(\text{Fe}, \text{Ni})_{0.75} : (\text{C}, \text{Va})_{0.25}$.

13.2 Derive the analogous equations to (13.16) and (13.17) for component A.

13.3 Derive integral and partial expressions for the mixing entropy in a four-sublattice model for the most ordered and random solutions. Provide plots of the results.

13.4 Titanium monocarbide possesses the NaCl structure but, unlike NaCl, TiC exists over a wide range of homogeneity, varying from 32 mol % C to a maximum of 48.8 mol % C. using the sublattice notation, write down the appropriate model for this phase and calculate the Bragg–Williams entropy of mixing of this carbide having 46 mol % C. Give your results in both joules per mole of atoms and per mole of Ti.

14 Solution Phase Models II: Configurational Energy

In the previous chapter we introduced a factorization of the Gibbs energy in order to consider a basic thermodynamic model for a substitutional alloy phase. In this model, the experimental (or calculated) results for the Gibbs energy of formation of an alloy are assumed to contain separable configuration-dependent and configuration-independent contributions (the rationale behind this assumption is discussed in chapter 16):

$$\Delta_{\text{mix}} G_m = \Delta_{\text{mix}} H_m^{\text{conf}} - T \, \Delta_{\text{mix}} S_m^{\text{conf}} + \Delta_{\text{mix}} G_m^{\text{conf indep}}$$

We then calculated the configurational mixing entropy for this model for two extremes of a sublattice phase alloy—the most ordered state and the random state. Under these circumstances we could use the Boltzmann equation.

In this chapter, we turn our attention to the enthalpy of mixing, $\Delta_{\text{mix}} H_m^{\text{conf}}$. We will also make the additional assumption that the configurational part of the energy is *volume independent* so that $\Delta_{\text{mix}} H_m^{\text{conf}}$ is equal to the energy of mixing, $\Delta_{\text{mix}} U_m^{\text{conf}}$.

The aim is to express $\Delta_{\text{mix}} U_m^{\text{conf}}$ in a form which depends on the atomic distribution, that is, in terms of local environment effects. In the simplest type of ordered alloy, comprised of two sublattices, the nearest neighbors of A atoms are the maximum number possible of B atoms. In some lattices, however, the nearest neighbors in the fully ordered state cannot all be B atoms. In an fcc alloy, for example, it is not possible to arrange for all nearest neighbors to be unlike atoms and obtain an ordered structure. This is called *frustration*. In a disordered alloy there will be a significant proportion of like-atom neighbors with the actual ratio of like and unlike depending on the temperature and composition and the degree of short-range order (SRO).

In almost all thermodynamic models of solid solution phases, the experimental values for $\Delta_{\text{mix}} H_m^{\text{conf}}$ are mapped onto the energy parameters of what is called an *Ising-like model*. The original Ising model was for a lattice model of a ferromagnet in which the atomic magnetic moments (designated as up-spin or down-spin) were located on the lattice sites and interacted via constant pairwise exchange energies. There is a one-to-one correspondence between this model of a magnet

Materials Thermodynamics. By Y. Austin Chang and W. Alan Oates
Copyright 2010 John Wiley & Sons, Inc.

and that for a binary alloy, with the up-spin being equivalent to an A atom and a down-spin to a B atom. In order to extend the model by including the possibility of energy representations other than only pair representations, the term Ising-like model is used.

In using an Ising-like model, it is usually assumed that *the model energy parameters are configuration independent*; that is, we can apply the energy parameters derived from one configuration to other configurations in that alloy. We may, for example, use the information on $\Delta_{mix}U_m^{conf}$ for the fully ordered stoichiometric compound to calculate the Ising-like energy parameters and then use these same parameters in order to calculate $\Delta_{mix}U_m^{conf}$ for the partially ordered nonstoichiometric phase or for the disordered phase.

As will become apparent later, it should be emphasized that *there is no unique way of carrying out the energy mapping*. This should make it clear that the derived model energy parameters should not be confused with believing that they are actual chemical bonding energies. We are wanting to map the total energy in such a way that it embraces only the near neighbors; that is, we are assuming that the interactions outside a small circle of influence are substantially weaker than those inside the circle.

14.1 PAIR INTERACTION MODEL

In order to illustrate, we will consider two examples. In the first, only one ordered phase is formed, at the composition AB. The configurational energy in the BW most ordered state will vary with composition as shown in Figure 14.1.

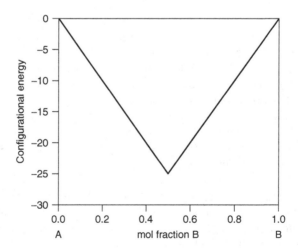

Figure 14.1 Configurational mixing energy for BW most ordered alloys in two-sublattice model.

We consider a model comprising two sublattices only and assume that these are of equal size. This is the same model considered in the previous chapters for the calculation of the configurational mixing entropies and can be represented by the notation

$$(A,B)_{0.5} : (B,A)_{0.5}$$

With two sublattices of equal size there can only be one fully ordered stoichiometric compound, AB. At 0 K for this stoichiometric alloy, all the A atoms are on one sublattice (1) and all the B atoms on the other (2). At higher temperatures, or with deviations from stoichiometry, antisite disorder excitation of the atoms occurs (some of the atoms occupy the wrong sublattice). Eventually, at a sufficiently high temperature or deviation from stoichiometry, the A and B atoms become evenly distributed on both sublattices. At this stage there is only one lattice; that is, we have a substitutional solid solution. The phase has undergone an order/disorder transformation.

14.1.1 Ground-State Structures

The term ground-state structure is used to refer to the most stable structure at 0 K. Even with just the two sublattices of equal size, it is possible to obtain different structures at the stoichiometric composition AB. In the structures based on the bcc lattice, two possibilities are shown in Figure 14.2—the occupation of the two sublattices is different in the two cases. It is clear, therefore, that in our thermodynamic modeling, it is not sufficient to know only the stoichiometric composition of the phase in order to be able to map the configurational energy onto an Ising-like model. We must also know the *ground-state structure* of the alloy. It is obvious that, in Figure 14.2a, the B2 structure will be stable when the nearest neighbor pair interaction energy between unlike atoms, $\varepsilon_{AB}^{(1)} < 0$ dominates, while the B32 structure shown in Figure 14.2b will not be stable under these conditions. Rather, it will be necessary to have $\varepsilon_{AB}^{(2)} < 0$ and with this second nearest neighbor interaction dominating in order for this

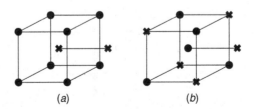

(a)   (b)

Figure 14.2   (a) the B2 and (b) B32 ground-state structures in a bcc lattice. The dots represent the A atoms and the crosses the B atoms.

structure to become stable. But over what range of $\varepsilon_{AB}^{(2)}/\varepsilon_{AB}^{(1)}$ will the B2 structure be stable? To answer this question, we must evaluate which structure has the lower formation energy over all possible values of the ratio. The results from this kind of study can then be summarized in a *phase stability map*. Such maps, showing the various possible ground-state structures for different ratios of the pair interaction energies, are available for the fcc, hcp, and bcc lattices.

We will simplify our two-sublattice model by considering nearest neighbor interaction mappings only. In the case of a bcc-based intermediate phase, this means that the compound AB has the B2 structure. As we have indicated, this phase would remain stable if we took $\varepsilon_{AB}^{(2)}/\varepsilon_{AB}^{(1)}$ to be other than zero. Different values of $\varepsilon_{AB}^{(1)}$ and $\varepsilon_{AB}^{(2)}$ could be selected to agree with the same $\Delta_{\mathrm{mix}}U_m^{\mathrm{conf}}$ values, a result which emphasizes that there is no unique mapping of the configurational energy onto an Ising-like model's energy parameters.

14.1.2 Nearest Neighbor Model

As shown in Figure 14.3 for the two-dimensional equivalent lattice for the sublattice model being discussed, nearest neighbors reside on different sublattices. Examples of the four different types of inter-sublattice pairs in a binary alloy are shown. In our model we assume that the alloy's configurational energy is being mapped onto these nearest neighbor pair interactions only.

The molar configurational energy of the alloy, in any given atomic arrangement, can be expressed in terms of the average probabilities of the different types of pairs and the pair energies, ε_{PQ}, whose values are to be determined. For a binary AB system, the alloy energy is

$$U_m^{\mathrm{alloy}} = \langle p_{AA}\rangle\varepsilon_{AA} + \langle p_{AB}\rangle\varepsilon_{AB} + \langle p_{BA}\rangle\varepsilon_{BA} + \langle p_{BB}\rangle\varepsilon_{BB} \tag{14.1}$$

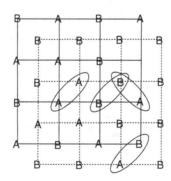

Figure 14.3 Two-dimensional two-sublattice model in which it is assumed that only nearest neighbor interactions are present.

where the angular bracketed terms refer to the mean probability of occurrence of a particular type of pair.

Note that, since the dimensions of U_m are per mole of sites, then the dimensions of the pair interaction energies used in (14.1) also refer to an energy per mole of sites and not, for the moment, to an energy per mole of pairs.

For the mechanical mixture (or the fully segregated alloy or the composition weighted sum) of the pure components *in the same structure as that of the intermediate phase*,

$$U_m^{\text{mixt}} = x_A \varepsilon_{AA} + x_B \varepsilon_{BB} \tag{14.2}$$

The equations which relate the mole fractions to the average pair probabilities are

$$x_A = \langle p_{AA} \rangle + \tfrac{1}{2}(\langle p_{AB} \rangle + \langle p_{BA} \rangle)$$

$$x_B = \langle p_{BB} \rangle + \tfrac{1}{2}(\langle p_{AB} \rangle + \langle p_{BA} \rangle)$$

Note that, when these two equations are added, the sum of both the left-hand and right-hand sides equals 1, as they should.

We can now rewrite the equation for the mechanical mixture as

$$U_m^{\text{mixt}} = \left[\langle p_{AA} \rangle + \tfrac{1}{2} \left(\langle p_{AB} \rangle + \langle p_{BA} \rangle \right) \right] \varepsilon_{AA}$$
$$+ \left[\langle p_{BB} \rangle + \tfrac{1}{2} \left(\langle p_{AB} \rangle + \langle p_{BA} \rangle \right) \right] \varepsilon_{BB} \tag{14.3}$$

Subtracting (14.3) from (14.1) gives the mixing energy:

$$\Delta_{\text{mix}} U_m^{\text{conf}} = \langle p_{AB} \rangle \left[\varepsilon_{AB} - \tfrac{1}{2}(\varepsilon_{AA} + \varepsilon_{BB}) \right]$$
$$+ \langle p_{BA} \rangle \left[\varepsilon_{BA} - \tfrac{1}{2}(\varepsilon_{AA} + \varepsilon_{BB}) \right]$$

or

$$\boxed{\Delta_{\text{mix}} U_m^{\text{conf}} = \tfrac{1}{2} z \left(\langle p_{AB} \rangle + \langle p_{BA} \rangle \right) W_{AB}} \tag{14.4}$$

where $W_{AB} = \varepsilon_{AB} - 1/2(\varepsilon_{AA} + \varepsilon_{BB})$ is called the *pair exchange energy* and we have taken $W_{AB} = W_{BA}$; that is, the pair exchange energy is assumed independent of the actual location of the A and B atoms.

Note that W_{AB} is now expressed as the energy per mole of pairs by multiplying the previous energies per mole of sites by the number of pairs per site ($= z/2$, where z is the coordination number of the lattice; $z = 4$ in Fig. 14.3).

As we did with the mixing entropy, we can evaluate (14.4) for the two limiting cases considered in Chapter 13, namely, the BW most ordered and the random solution.

The pair probability distributions for the BW most ordered state are shown in Figure 14.4a, where it can be seen that

$$\langle p_{BA} \rangle = 0 \qquad\qquad \text{all compositions}$$

$$\langle p_{AB} \rangle = \begin{cases} 2x_B & \text{for } x_B \leq \tfrac{1}{2} \\ 2(1 - x_B) & \text{for } x_B > \tfrac{1}{2} \end{cases}$$

$$\langle p_{AA} \rangle = \begin{cases} 1 - 2x_B & \text{for } x_B \leq \tfrac{1}{2} \\ 0 & \text{for } x_B > \tfrac{1}{2} \end{cases}$$

$$\langle p_{BB} \rangle = \begin{cases} 0 & \text{for } x_B \leq \tfrac{1}{2} \\ 2x_B - 1 & \text{for } x_B > \tfrac{1}{2} \end{cases}$$

For the random solution case, we can use the binomial expansion to calculate the probabilities:

$$\langle p_{AB} \rangle = \langle p_{BA} \rangle = \langle p_A \rangle \langle p_B \rangle = x_A x_B$$

$$\langle p_{AA} \rangle = x_A^2 \qquad \langle p_{BB} \rangle = x_B^2$$

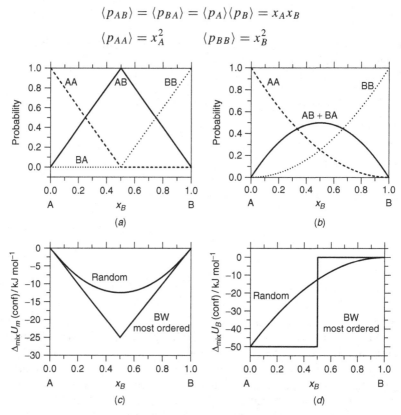

Figure 14.4 Pair probabilities and integral and partial molar configurational mixing energies for BW most ordered and random states in two-sublattice model: (a) pair probabilities for BW most ordered state; (b) pair probabilities for random state; (c) $\Delta_{mix}U_m$ for BW most ordered and random states; (d) $\Delta_{mix}U_B$ for BW \newline most ordered and random states.

The pair probability distributions for the random state are shown in Figure 14.4b.

Substitution of these equations for the two limiting cases into (14.4) gives the following mixing energies:

$$\Delta_{\text{mix}} U_m^{\text{conf}} = \begin{cases} z W_{AB} x_B & \text{BW most ordered,} & x_B \leq \frac{1}{2} \\ z W_{AB} x_A & \text{BW most ordered,} & x_B > \frac{1}{2} \\ z W_{AB} x_A x_B & \text{random solution, all compositions} \end{cases}$$

The partial molar properties for the most ordered and random mixing limits may now be obtained from these integral quantities. For the component B,

$$\Delta_{\text{mix}} U_B^{\text{conf}} = \begin{cases} z W_{AB} & \text{BW most ordered,} & x_B \leq \frac{1}{2} \\ 0 & \text{BW most ordered,} & x_B > \frac{1}{2} \\ z W_{AB} x_A^2 & \text{random solution, all compositions} \end{cases}$$

It can be seen that the mixing energy for the disordered (random) state is of exactly the same form as that already met when using the Redlich–Kister representation with $L_0 = z W_{AB}$. Thus we have *one* possible model which physically models the behavior of this empirical modeling equation. It would be wrong, however, to believe that the *only* model to yield this result is this nearest neighbor pair interaction model.

Both $\Delta_{\text{mix}} U_m^{\text{conf}}$ and $\Delta_{\text{mix}} U_B^{\text{conf}}$ for these two limiting cases are shown in Figures 14.4c and 14.4d. It can be clearly seen that, as the temperature is increased in going from the most ordered state to the random disordered state, there is a decrease in the magnitude of $\Delta_{\text{mix}} U_m^{\text{conf}}$, even though we have used the same temperature- (and composition-) independent energy parameter in this Ising-like model.

Note the physical inconsistency in our having used a model which possesses a finite mixing energy when we have assumed a random mixing entropy. Finite pair interaction energies will give rise to SRO, which will in turn lead to both a modified mixing energy and entropy. The BW model for sublattice phases and the point, zeroth, or regular solution approximation for substitutional solid solutions ignores this fact.

14.2 CLUSTER MODEL

The two-sublattice model discussed above can be applied to ordered phases whose stoichiometry is different from $f^\alpha = f^\beta = \frac{1}{2}$. Similar arguments and derivations to those used above could be equally well applied to, for example, A_3B, but, in every case, $\Delta_{\text{mix}} U_m^{\text{conf}}$ for the most ordered state will look similar to that shown in Figure 14.1, that is, with a single sharp minimum at the stoichiometric composition.

Two types of complication can arise, however, which suggest that improvements to our two-sublattice model may sometimes be necessary:

(a) If structure is to be taken into consideration, it may be necessary to introduce interactions between more distant neighbors than just the nearest. The case of bcc alloys was mentioned earlier—the B2 structure is expected when $\varepsilon_{AB}^{(2)}/\varepsilon_{AB}^{(1)}$ is small but the B32 structure is expected to be stable when it is large.

(b) Ordered phases can form at more than one composition from the same parent disordered phase. In this case $\Delta_{mix}U_m^{conf}$ for the most ordered state will look like that shown in Figure 14.5. Here, there are three ordered phases to consider, A_3B, AB, and AB_3. It can be seen that there is only one minimum in this curve, with inflexions in the curve occurring at the other ordered phase compositions. The classic example of this kind of behavior is found in the Au–Cu system where the disordered fcc parent phase gives rise to ordered phases at the Au_3Cu, $CuAu$, and Cu_3Au compositions.

In order to take these points into consideration we will use a four-sublattice model instead of our previous two-sublattice model. The notation for this sublattice model is

$$(A, B)_{0.25} \;:\; (A, B)_{0.25} \;:\; (A, B)_{0.25} \;:\; (A, B)_{0.25}$$

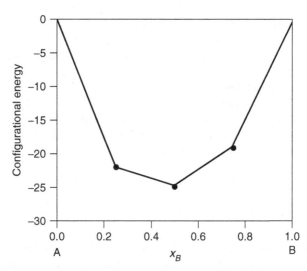

Figure 14.5   Configurational mixing energy for fully ordered alloys in four-sublattice model.

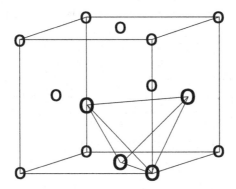

Figure 14.6   Nearest neighbor tetrahedron cluster in fcc lattice.

The four near neighbor sites form a tetrahedron cluster in both the fcc and bcc structures. In the fcc lattice the six tetrahedron edges are the nearest neighbors, as illustrated in Figure 14.6. In the bcc lattice, on the other hand, the nearest and next nearest neighbors form an irregular tetrahedron, as is apparent from Figure 14.2.

In the cluster model, the alloy's configurational energy is given by

$$U_m^{\text{alloy}} = \sum_{i=1}^{5} p_i \varepsilon_i \qquad (14.5)$$

where the ε_i are the cluster energies per site.

Just as in the case of the pair approximation, where we used $W_{AB} = W_{BA}$, we will assume that the energy of a cluster depends only on the *number* of the various participating atoms and not on their *arrangement*, for example, $\varepsilon_{ABBA} = \varepsilon_{BAAB} = \cdots = \varepsilon_{A_2B_2}$. This means that, in a binary system with a 4-point cluster, there are only 5 independent alloy cluster energies to consider instead of the 16, were we to consider the influence of atom arrangements on the cluster energies as well.

As we did for the pair approximation model, we can write the equation for the mechanical mixture:

$$U_m^{\text{mixt}} = x_A \varepsilon_{A_4} + x_B \varepsilon_{B_4} \qquad (14.6)$$

and use the relation between the mole fractions and the cluster probabilities (note that both sides sum to unity):

$$x_A = \langle p_{A_4} \rangle + \tfrac{3}{4}\langle p_{A_3B} \rangle + \tfrac{1}{2}\langle p_{A_2B_2} \rangle + \tfrac{1}{4}\langle p_{AB_3} \rangle \qquad (14.7)$$

$$x_B = \tfrac{1}{4}\langle p_{A_3B} \rangle + \tfrac{1}{2}\langle p_{A_2B_2} \rangle + \tfrac{3}{4}\langle p_{AB_3} \rangle + \langle p_{B_4} \rangle \qquad (14.8)$$

The cluster probabilities for the most ordered and the random states are shown in Figures 14.7a and 14.7b, respectively.

In the case of the pair cluster in the two-sublattice model, there was no possibility of the energy being shared between the clusters. This is not the case, however, in the case of tetrahedron clusters in the fcc lattice. It is possible to closely pack the tetrahedra so that they share edges or to loosely pack them in a different way so that there is no sharing of edges, but only of corners. In the two cases, the number of clusters per site, γ, and the energy assigned to a single cluster must differ but the resulting total energy must be the same in both cases.

The mixing energy is given by the difference of (14.5) and (14.3) while using (14.7):

$$\Delta_{mix} U_m^{conf} = \gamma [\langle p_{A_3B} \rangle V_{A_3B} + \langle p_{A_2B_2} \rangle V_{A_2B_2} + \langle p_{AB_3} \rangle V_{AB_3}] \qquad (14.9)$$

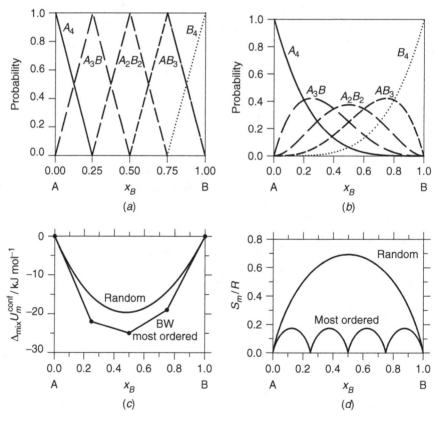

Figure 14.7 Cluster probabilities and configurational mixing energies and entropies for BW most ordered and random states in four-sublattice model: (*a*) cluster fractions in BW most ordered state; (*b*) cluster fractions in random state; (*c*) $\Delta_{mix} U_m$ for BW most ordered and random states; (*d*) $\Delta_{mix} S_m$ for BW most ordered and random states.

where $V_{AB_3} = \varepsilon_{AB_3} - (1/4\varepsilon_{A_4} + 3/4\varepsilon_{B_4})$, and so on. These cluster exchange energies are analogous to the pair exchange energies in the pair approximation used previously. Because of the introduction of the γ conversion factor (the number of clusters per site), the cluster energies are now in units of energy per cluster.

We can use the results for the most ordered and for the random states shown in Figures 14.7a and 14.7b together with (14.9) to obtain $\Delta_{mix}U_m^{conf}$ for these two limiting cases. For the most ordered case, the values between the stoichiometric compounds are simply the linear joins, while, for the random solution case, $\Delta_{mix}U_m^{conf}$ is obtained from a binomial expansion and is given by

$$\Delta_{mix}U_m^{conf} = \gamma(4x_A^3 x_B V_{A_3B} + 6x_A^2 x_B^2 V_{A_2B_2} + 4x_A x_B^3 V_{AB_3}) \tag{14.10}$$

The calculated integral energies and entropies of mixing for the most ordered and random alloys are shown in Figures 14.7c and 14.7d, respectively.

If desired, it is possible to continue to express the cluster energies in terms of the near neighbor pair interaction energies. For the fcc-based lattice and assuming that W_{AB} is constant, it is easy to see that, at the AB composition, $\Delta_{mix}U_m^{conf} = 4W_{AB}$, whereas at the A_3B and AB_3 compositions, $\Delta_{mix}U_m^{conf} = 3W_{AB}$.

However, if we use the values for $\Delta_{mix}U_m^{conf}$ given in Figure 14.5, we would find that there are three different values for W_{AB}:

$$A_3B: \qquad W_{AB} = \frac{-22}{3} = -7.33 \text{ kJ}$$

$$AB: \qquad W_{AB} = \frac{-25}{4} = -6.25 \text{ kJ}$$

$$AB_3: \qquad W_{AB} = \frac{-19}{3} = -6.33 \text{ kJ}$$

In this situation, we might accept the volume (or composition) dependence of W_{AB} or, alternatively, use the cluster energies.

Summarizing, any energy representation adopted should give rise to the observed ground-state structures for that system.

The models discussed have used the following important *assumptions*:

- Additive separation of the Gibbs energy into configuration-dependent and configuration independent contributions.
- The configuration-dependent energy is volume independent. This may be the weakest of the assumptions used.
- The cluster energy parameters are temperature and composition independent.
- The parameters of the Ising-like model can be used for all configurations.
- The energies of clusters, including pairs, depend only on the number and not on the arrangement of the different types of atoms, for example, $V_{ABBA} = V_{BAAB} = V_{A_2B_2}$, $W_{AB} = W_{BA}$.

EXERCISES

14.1   Derive the expressions for the partial molar energy of B in the tetrahedron cluster model for the most ordered and random mixing cases. Using the integral mixing energy and entropy values given in Figures 14.7c and 14.7d, calculate the partial molar values for both properties in the most ordered and random states.

15 Solution Models III: The Configurational Free Energy

In the last two chapters, we have seen how:

1. The configurational entropy for a *sublattice phase* can readily be calculated in the point, zeroth, or BW approximation (hereafter referred to as the BW approximation) for any set of sublattice mole fractions, $y_i^{(j)}$. In this approximation, the atoms on the individual sublattices are assumed to be randomly distributed and the configurational mixing entropy $\Delta_{\mathrm{mix}} S_m^{\mathrm{conf}}$ is given by

$$\Delta_{\mathrm{mix}} S_m^{\mathrm{conf}} = -R \sum_{(j)} f^{(j)} \sum_i y_i^{(j)} \log_e y_i^{(j)} \qquad (15.1)$$

2. The total configurational energy of the phase can be mapped onto the energy parameters of an Ising-like model. In terms of a four-point cluster model, the energy is related to the cluster probabilities and the cluster energies by

$$\Delta_{\mathrm{mix}} U_m^{\mathrm{conf}} = \gamma \sum_{\alpha\beta\gamma\delta} \sum_{ijkl} \langle p_{ijkl}^{\alpha\beta\gamma\delta} \rangle V_{ijkl} \qquad (15.2)$$

where the superscripts on the V's have been dropped because of our assumption that the cluster energy depends only on the number or the different types of atoms in the cluster and not on their arrangement.

In the previous chapter we used (15.1) and (15.2) to calculate the properties for two extreme cases—the BW most ordered state and the the random mixing state, corresponding with 0 K and infinite temperature, respectively. We now wish to calculate the configurational Helmholtz energy of mixing, $\Delta_{\mathrm{mix}} A_m^{\mathrm{conf}}$, for any given temperature and composition. Since our configurational model is a constant-volume model, the appropriate free energy is the Helmholtz energy and not the Gibbs energy.

The calculation is simplified by using the BW approximation where, no matter how large the magnitude of the pair or cluster energies, the atoms are assumed to be distributed randomly on the individual sublattices. Similarly, in

Materials Thermodynamics. By Y. Austin Chang and W. Alan Oates
Copyright 2010 John Wiley & Sons, Inc.

the single-lattice disordered phase, we assume that the distribution of atoms on that lattice is also independent of the magnitude of the pair or cluster formation energies. This BW approximation is clearly not a self-consistent one since a set of large negative pair or cluster formation energies means a strong tendency for nonrandom mixing. The approach is very useful, however, for illustrating solution-phase modeling. It has the big advantage that the energy and entropy may be treated separately. When we allow for the coupling between the energy and entropy, it is necessary to use more sophisticated models.

The assumed random mixing on the sublattices means that the cluster probabilities are given by the product of the average site occupation probabilities on the individual sublattices which, in turn, are the same as the sublattice mole fractions. For a four-point cluster

$$\langle p_{ijkl}^{\alpha\beta\gamma\delta} \rangle = \langle p_i^{\alpha} \rangle \langle p_j^{\beta} \rangle \langle p_k^{\gamma} \rangle \langle p_l^{\delta} \rangle = y_i^{\alpha} y_j^{\beta} y_k^{\gamma} y_l^{\delta}$$

In this approximation, $\Delta_{\text{mix}} U_m^{\text{conf}}$ is given by

$$\Delta_{\text{mix}} U_m^{\text{conf}} = \gamma \sum_{\alpha\beta\gamma\delta} \sum_{ijkl} y_i^{\alpha} y_j^{\beta} y_k^{\gamma} y_l^{\delta} V_{ijkl} \tag{15.3}$$

and the equation for the *configurational* Helmholtz energy of mixing in this cluster model in the BW approximation becomes

$$\Delta_{\text{mix}} A_m^{\text{conf}} = \gamma \sum_{ijkl} \sum_{ijkl} y_i^{\alpha} y_j^{\beta} y_k^{\gamma} y_l^{\delta} V_{ijkl}$$

$$+ RT \sum_{(j)} f^{(j)} \sum_i y_i^{(j)} \log_e y_i^{(j)} \tag{15.4}$$

It should be noted that, by using the BW approximation, the 16 tetrahedron probabilities, $p_{ijkl}^{\alpha\beta\gamma\delta}$, have been reduced to only 4 point probabilities, y_i^{α}. This reduction is extremely important from a computational viewpoint.

Equation (15.4) does not contain the mole fractions as a variable so that we cannot simply substitute values for T and x_B and obtain a value of $\Delta_{\text{mix}} A_m^{\text{conf}}$, as we are able to do when using a polynomial representations of $\Delta_{\text{mix}} A_m^{\text{conf}}$ values.

15.1 HELMHOLTZ ENERGY MINIMIZATION

The $y_i^{(j)}$'s are *internal variables* in the minimization of $\Delta_{\text{mix}} A_m^{\text{conf}}$; that is, it is a *functional* rather than a function. Equilibrium at any composition depends on minimizing $\Delta_{\text{mix}} A_m^{\text{conf}}$ with the y_i^{α} as the independent variables but subject to the constraint of constant composition (recall that $x_i = \sum_{(j)} \sum_i f^{(j)} y_i^{\alpha}$).

In order to discuss the procedure and the kind of results obtained, we will consider the nearest neighbor two-sublattice model and assume that the single energy parameter W_{AB} has been obtained from a knowledge of the *configura-tional* energy for a known atom distribution. Usually, this is from information on the most ordered state. In the following we assume that W_{AB} is temperature and composition independent.

For two equally sized sublattices in the pair approximation, (15.4) can be written out in full as

$$\Delta_{mix} A_m^{conf} = \tfrac{1}{2} z [y_A^\alpha y_B^\beta + y_B^\alpha y_A^\beta] W_{AB}$$
$$+ \tfrac{1}{2} RT [y_A^\alpha \log_e y_A^\alpha + y_B^\alpha \log_e y_B^\alpha + y_A^\beta \log_e y_A^\beta + y_B^\beta \log_e y_B^\beta] \tag{15.5}$$

The low- and high-temperature limits of (15.5) can be obtained from equations given in the two previous chapters. In the high-temperature limit (the random state)

$$\Delta_{mix} A_m^{conf} = z x_A x_B W_{AB} + RT (x_A \log_e x_A + x_A \log_e x_B)$$

and in the low-temperature limit (the most ordered state) and for $x_B < \tfrac{1}{2}$

$$\Delta_{mix} A_m^{conf} = z x_B W_{AB} + RT [(1 - 2x_B) \log_e (1 - 2x_B) + 2x_B \log_e (2x_B)]$$

We can also minimize the Helmholtz energy functional, (15.5), for intermediate stages of ordering and obtain results for the various properties which lie between those for the the low- and high-temperature limits.

The Lagrangian multiplier method could be used to solve the constrained minimization problem, but a simpler way, in this particular case, is to change the variables so that the constraining conditions are built into the functional. To do this, we define a long-range order (LRO) parameter as follows:

$$\eta = y_A^\alpha - y_A^\beta = y_B^\beta - y_B^\alpha$$

In the disordered state, where $y_B^\alpha = y_B^\beta$, $\eta = 0$. In the ordered state, where $y_B^\alpha \neq y_B^\beta$ ($y_B^\beta > y_B^\alpha$), $\eta > 0$.

To change variables, we use the constraining condition on the sublattice mole fractions:

$$x_A = \tfrac{1}{2} (y_A^\alpha + y_A^\beta) \qquad x_B = \tfrac{1}{2} (y_B^\alpha + y_B^\beta)$$

and then express the y_i^α's in terms of the x_i's and η:

$$\begin{array}{ll} y_A^\alpha = x_A + \tfrac{1}{2} \eta & y_A^\beta = x_A - \tfrac{1}{2} \eta \\ y_B^\alpha = x_B - \tfrac{1}{2} \eta & y_B^\beta = x_B + \tfrac{1}{2} \eta \end{array} \tag{15.6}$$

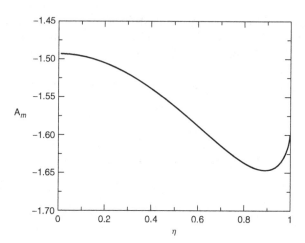

Figure 15.1 Variation of Helmholtz energy with long-range parameter η for $zW_{AB}/RT = -3.2$ at $x_B = 0.5$.

If we now insert (15.6) into (15.3) and (15.1), we can remove the sublattice mole fractions from the equations for the mixing quantities:

$$\Delta_{mix} U_m^{conf} = \tfrac{1}{2} z W_{AB} \left(2 x_A x_B + \tfrac{1}{2}\eta^2\right) \tag{15.7}$$

$$\Delta_{mix} S_m^{conf} = -\tfrac{1}{2} R[(x_A + \tfrac{1}{2}\eta) \log_e(x_A + \tfrac{1}{2}\eta) + (x_B - \tfrac{1}{2}\eta) \log_e(x_B - \tfrac{1}{2}\eta)$$
$$+ (x_A - \tfrac{1}{2}\eta) \log_e(x_A - \tfrac{1}{2}\eta) + (x_B + \tfrac{1}{2}\eta) \log_e(x_B + \tfrac{1}{2}\eta)] \tag{15.8}$$

so that we now have $\Delta_{mix} A_m^{conf}$ as a function of the single internal variable, η, with the constraints for the relation between lattice and sublattice mole fractions already incorporated. An example showing how A_m varies with η is given in Figure 15.1. It can be seen that, at the particular value of W_{AB}/RT chosen, there is a Helmholtz minimum close to $\eta = 0.9$. Once solved for the equilibrium value of η, (15.6) can be used to obtain the sublattice mole fractions.

In Chapter 13, we showed how the sublattice mole fractions varied with the mole fraction for the most ordered case. Figure 15.2 shows this variation at intermediate levels of the disordering.

Minimization of $\Delta_{mix} A_m^{conf} (= \Delta_{mix} U_m^{conf} - T \Delta_{mix} S_m^{conf})$ with respect to η yields

$$\left(\frac{\partial \Delta_{mix} A_m}{\partial \eta}\right)_{x_B} = \eta \frac{z}{2} W_{AB} + \frac{RT}{4} \log_e \left[\frac{(x_A + \eta/2)(x_B + \eta/2)}{(x_A - \eta/2)(x_B - \eta/2)}\right] = 0 \tag{15.9}$$

The sudden changes in the slopes of the integral quantity curves at compositions which depend on temperature is due to the change from the disordered phase (single lattice) to the ordered phase (two sublattices). In this particular case of a second-order order/disorder transformation, there is no discontinuity

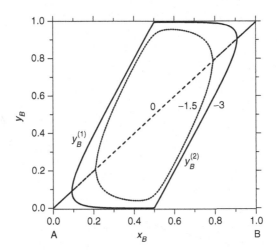

Figure 15.2 Sublattice mole fractions of B for different values of W_{AB}/RT (numbers on the curves) in the two-sublattice model.

(only a *singularity*) in the first derivatives of $\Delta_{\text{mix}} A_m$(conf) (both $\Delta_{\text{mix}} S_m^{\text{conf}} = -\partial \Delta_{\text{mix}} A_m / \partial T$ and $\Delta_{\text{mix}} \mu_i^{\text{conf}} = -\partial \Delta_{\text{mix}} A / \partial N_i$ at the transition). There are, however, *discontinuities* in the second derivatives at the phase boundary and these can be seen in Figure 15.3 for $\Delta_{\text{mix}} S_B^{\text{conf}}$ and $\Delta_{\text{mix}} H_B^{\text{conf}}$.

It is clear from these plots why a polynomial representation of the thermodynamic properties as a function of composition and temperature is out of the question for this type of intermediate phase.

15.2   CRITICAL TEMPERATURE FOR ORDER/DISORDER

The temperature at which the distinction between the sublattices, or LRO, disappears is the order/disorder critical temperature. In the case of the pair cluster, two-sublattice model in the BW approximation, the limiting value of W_{AB} where $\eta \to 0$ can be obtained from (15.9) by concentrating on the critical composition, $x_A = \frac{1}{2}$:

$$\eta \frac{z}{2} W_{AB} + \frac{RT_c \eta}{4 x_A x_B} = 0 \qquad (15.10)$$

where we have used the general relation

$$\lim \delta \to 0 \qquad \log_e \left[\frac{x + \delta}{x - \delta} \right] = \frac{2\delta}{x}$$

At the critical composition, (15.10) becomes

$$\boxed{\frac{z}{2} W_{AB} = -RT_c} \qquad \begin{array}{l} \text{BW approx.} \\ \text{order/disorder} \end{array} \qquad (15.11)$$

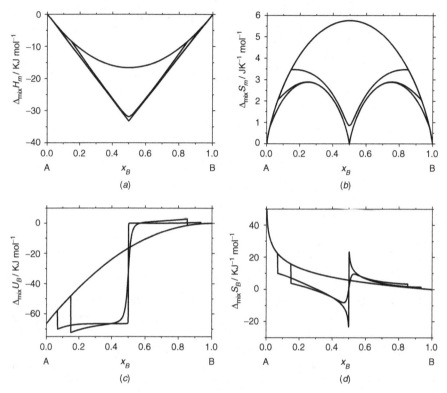

Figure 15.3 Integral and partial molar configurational thermodynamic properties for two-sublattice model for different values of W_{AB}/RT $(0, -1.5, -3)$: (a) $\Delta_{\text{mix}}U_m^{\text{conf}}$; (b) $\Delta_{\text{mix}}S_m^{\text{conf}}$; (c) $\Delta_{\text{mix}}U_B^{\text{conf}}$; (d) $\Delta_{\text{mix}}S_B^{\text{conf}}$. The values for the BW most ordered structures are also shown for the integral quantities.

Note that this relation for the order/disorder critical point is different from that for the segregation critical point. The latter can be obtained from putting $\partial^2 A_m/\partial x_B^2 = \partial^3 A_m/\partial x_B^3 = 0$ with $\eta = 0$ in (15.7) and (15.8). The result is

$$\frac{z}{4}W_{AB} = RT_c \quad \begin{array}{l} \text{BW approx.} \\ \text{phase separation} \end{array} \quad (15.12)$$

Figure 15.4a shows the resulting phase diagram for this two-sublattice, nearest neighbor pair interaction model in the BW approximation. Figure 15.4b shows how the sublattice mole fractions change with temperature for an alloy with composition $x_B = 0.5$. The loss of LRO with increasing temperature can be seen. Unlike the case for most phase diagrams, where the different phases are separated by a two-phase region, there is only a single phase boundary on this particular phase diagram, which separates the ordered from the disordered region. This indicates that the order/disorder transformation is of the second-order type.

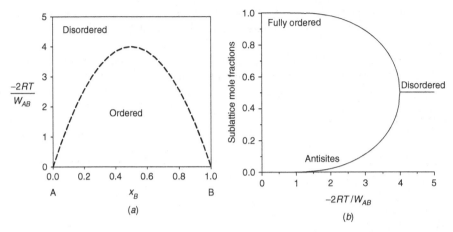

Figure 15.4 Two-sublattice, nearest neighbor pair interaction model in BW approxima-tion: (a) phase diagram; (b) sublattice mole fractions for $x_B = \frac{1}{2}$.

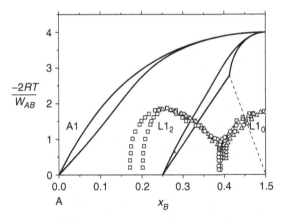

Figure 15.5 Comparison of phase diagram calculated using BW approximation with "exact" one obtained from Monte Carlo simulations for nearest neighbor pair interaction model in fcc alloys.

This is also apparent from Figure 15.4 and from a consideration of the integral and partial molar thermodynamic properties shown in Figures 15.3c and 15.3d.

Some other order/disorder transitions are found to be of first order (disconti-nuities are present in the first derivatives of A). There are then two-phase regions on the order/disorder phase diagram, as is illustrated in Figure 15.5. Both phase diagrams shown result from using nearest neighbor pair interaction energies in the four-sublattice model for the fcc lattice. The solid-line diagram is obtained by using the BW approximation (the dashed line is for a second-order transition between two ordered phases). The exact phase diagram obtained by using Monte

Carlo simulations is shown as individual points. It can be seen that the topology of the two diagrams are quite different as are the magnitudes of the RT/W_{AB} at the critical points and the absence of the second-order transition. This confirms, as we expect, that there is an inconsistency in using a model in which random sublattice mixing is assumed when there are finite interaction energies.

EXERCISES

15.1 Use (15.5) to derive expressions for the partial molar properties of the two-sublattice phase.

15.2 Derive (15.9).

15.3 Derive (15.10).

16 Solution Models IV: Total Gibbs Energy

It has been assumed earlier that $\Delta_{\mathrm{mix}}G_m$ can be separated additively as follows:

$$\Delta_{\mathrm{mix}}G_m = [\Delta_{\mathrm{mix}}H_m^{\mathrm{conf}} - T\,\Delta_{\mathrm{mix}}S_m^{\mathrm{conf}}] + \Delta_{\mathrm{mix}}G_m^{\mathrm{confindep}}$$

The configurational energy model used in Chapters 14 and 15 is a constant-volume model. The assumption of volume-independent pair or cluster energies is not a good one since a change of volume means a change in interatomic distances, which, in turn, will be expected to give rise to a change in interaction energies.

In a constant-volume model, $\Delta_{\mathrm{mix}}H_m^{\mathrm{conf}}$ can be replaced by $\Delta_{\mathrm{mix}}U_m^{\mathrm{conf}}$ so that

$$\Delta_{\mathrm{mix}}G_m = [\Delta_{\mathrm{mix}}U_m^{\mathrm{conf}} - T\,\Delta_{\mathrm{mix}}S_m^{\mathrm{conf}}] + \Delta_{\mathrm{mix}}G_m^{\mathrm{confindep}}$$

$$= \Delta_{\mathrm{mix}}A_m^{\mathrm{conf}} + \Delta_{\mathrm{mix}}G_m^{\mathrm{confindep}}$$

In this chapter we will concentrate on the term $\Delta_{\mathrm{mix}}G_m^{\mathrm{confindep}}$. We can demonstrate its importance by considering two examples:

1. The phase diagram for the Au–Ni system is shown in Figure 16.1, where it can be seen that a miscibility gap exists in the fcc solid solution. According to our simple configurational model, this would indicate a *repulsion* between Au and Ni atoms, ($W_{\mathrm{AuNi}} > 0$). But in the solid solution region, near to that marked by a cross on the diagram, it is known experimentally (from diffraction studies) that substantial short-range ordering exists, that is, that there is an *attraction* between Au and Ni atoms, ($W_{\mathrm{AuNi}} < 0$). This apparent discrepancy is due to our failure to take into account the large atomic size difference which exists between Au and Ni. All the models discussed to date have been constant-volume models.

2. We have seen that $\Delta_{\mathrm{mix}}S_m^{\mathrm{conf}}$ is always ≥ 0, as is illustrated in the upper part of Figure 16.2. In the lower part of this figure it is shown how $\Delta_f S_m$ varies with composition for the Al–Ni system. Although there are many different ordered phases with many different structures in this system, it is clear that there are other very important contributions to the total $\Delta_{\mathrm{mix}}G_m$ (changes of

Materials Thermodynamics. By Y. Austin Chang and W. Alan Oates
Copyright 2010 John Wiley & Sons, Inc.

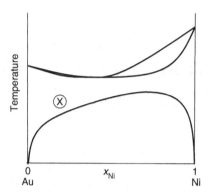

Figure 16.1 Phase diagram for Au–Ni system. Short-range order is known to exist in the region marked.

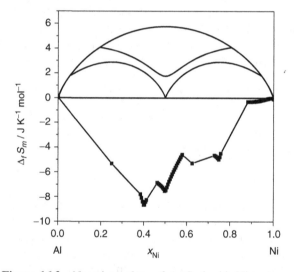

Figure 16.2 Negative values of $\Delta_f S_m$ in Al–Ni system.

reference states do not have a major effect). This same conclusion applies to many other alloy systems.

From these two examples it is clear that we should consider at least two contributions to $\Delta_{\text{mix}} G_m^{\text{confindep}}$:

$$\Delta_{\text{mix}} G_m^{\text{confindep}} = \Delta_{\text{mix}} G_m^{\text{asm}} + \Delta_{\text{mix}} G_m^{\text{excit}} \tag{16.1}$$

where the superscripts asm and excit refer to atomic size mismatch and thermal excitations, respectively.

Equation (16.1) implies the independence and separation of these contributing terms to $\Delta_{\text{mix}}G_m^{\text{confindep}}$ and remember that we have already assumed the independence of the configurational and nonconfigurational contributions. We will examine these assumptions later and discuss some important coupling effects which are being ignored when the terms are separated in this way.

16.1   ATOMIC SIZE MISMATCH CONTRIBUTIONS

The molar volumes of the pure components in a binary system are rarely the same and the alloys formed from these components will usually have intermediate molar volumes. This means that it is necessary to perform some work in bringing pure component A and pure component B to the volume of the alloy—in Figure 16.3 a dilation of A and the compression of B is necessary in the formation of the alloy. We are again simplifying by assuming that the metals and alloy are elastically isotropic.

The process being considered is sketched in Figure 16.3 and shown as a $p-V$ diagram in Figure 16.4. The $p-V_m$ relations for pure A, pure B and for 1 mol of an alloy $A_{1-x}B_x$ are shown. The energies involved for the individual components are shown shaded in Figure 16.4. The equilibrium volume V of the alloy is given by $\partial A/\partial V = 0$ and this corresponds with the $p = 0$ point on the curve for the alloy. Before the pure components are mixed and acquire the volume of the alloy, they are under compressive or dilatational stress. When they are now mixed to form the alloy, the pressure is again reduced to zero. Since there is no volume change in this stage of the process, there is no work done and no change in the Helmholtz energy.

The total deformation energy (Helmholtz energy) involved in the formation of 1 mol total of pure components in the amounts present in the alloy and at the

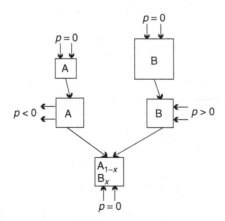

Figure 16.3   Molar volume changes on alloying.

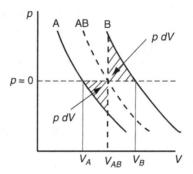

Figure 16.4 The $p - V$ work involved in alloying.

final volume of the alloy is

$$\Delta_{\text{mix}} A_m^{\text{asm}}(A_{1-x}B_x, V_{AB}) = (1 - x)\,\Delta A_m(A, V_A) + x\,\Delta A_m(B, V_B) \qquad (16.2)$$

The volume changes involved in this dilation/compression from the pure metal to the alloy are generally much larger than those found for the volume differences between the ordered and disordered forms of a given phase; that is, the assumption that the volume changes are composition independent is a fairly good one.

Consider first the dilation or compression of a pure solid. The work done *by* the system, $\delta w = p\,dV$, is negative in both cases (if $p > 0$, $dV < 0$ and $\delta W < 0$; if $p < 0$, $dV > 0$ and $\delta W < 0$).

The isothermal compressibility κ_T is defined by

$$\kappa_T = -\frac{1}{V^\circ}\left(\frac{\partial V}{\partial p}\right)_T$$

If it is assumed that κ_T is independent of p, as we did for a unary substance in Chapter 3, then integration of this equation gives

$$p = -\frac{V - V^\circ}{\kappa_T V^\circ}$$

so that the work done *on* the system increases the Helmholtz energy of the component by an amount

$$\Delta A = \int \frac{V - V^\circ}{\kappa_T V^\circ}\,dV \qquad (16.3)$$

so that we can write

$$\boxed{\Delta A_m = \frac{1}{2} V_m^\circ \left(\frac{\Delta V_m}{\kappa_T V_m^\circ}\right)^2} \qquad (16.4)$$

As expected, this equation demonstrates that there is a Helmholtz energy increase no matter whether the work done on the solid has been to compress it or dilate it.

We actually want the change in the Gibbs, as distinct from the Helmholtz, energy for this compress/dilate/mix process. Since

$$G = A + pV \qquad \Delta G = \Delta A + \Delta(pV)$$

then for the compress/dilate part of the process from the initial $p = 0$ state,

$$\Delta G = \Delta A + pV$$

and for the mix part of the process to the final $p = 0$ state,

$$\Delta G = 0 - pV$$

and we see that, for the compress/dilate/mix process, $\Delta_{mix} G_m = \Delta_{mix} A_m$.

Example 16.1 Calculation of Atomic Size Mismatch Contribution
Calculate $\Delta_{mix} G_m^{asm}$ for the Cr–Ta system.

The following table gives the property values at room temperature. We will simplify the calculation by making the (rather drastic) assumption that the molar density and elastic properties of Cr and Ta are the same in the alloy as in the pure components.

	V_m/cm^3 mol^{-1}	B_T/GPa
Cr	7.26	167
Ta	10.86	196

If we consider the alloy with $x = x_{Ta} = \frac{1}{2}$ and substitute the given property values into (16.4),

$$\Delta_{mix} G_m (Cr_{0.5} Ta_{0.5}, V) = \frac{(1/8) \times (10.86 - 7.26)^2 \times 10^{-12}}{(1/2) \left[7.26/167 + 16.86/196 \right] \times 10^{15}}$$

$$= +32.77 \text{ kJ (mol alloy)}^{-1}$$

If this calculation is repeated at different compositions, then the curve shown in Figure 16.5 is obtained.

In this particular system, the atomic size mismatch is quite large and, as a consequence, the effect is seen to give a very substantial contribution $\Delta_{mix} G_m$. As can be seen in Figure 16.5, the equation is approximately parabolic and equivalent to a value of $\Delta_{mix} G_m \approx 131 x_{Cr} x_{Ta}$ kJ mol^{-1} (much larger than the values considered when we examined some strictly regular solution-phase diagrams).

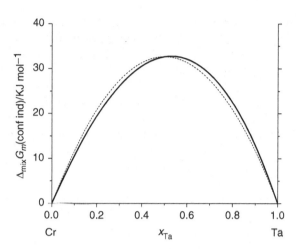

Figure 16.5 Calculated atomic size mismatch contribution, $\Delta_{\text{mix}}G_m^{\text{asm}}$, for (Cr,Ta) alloys. The dashed curve is a parabolic approximation to the curve calculated from (16.4).

It is clear from this example that, even though the model adopted is very crude, the atomic size mismatch contribution to the mixing properties, which is always positive, cannot be neglected. This configuration-independent contribution applies equally to ordered phases as well as to disordered phases; that is, this effect is due *not* to local elastic distortions of the atoms from their lattice sites but to the different molar volumes of the alloy as compared with those of the pure components. The effect of local lattice distortions is relatively small in comparison.

16.2 CONTRIBUTIONS FROM THERMAL EXCITATIONS

In the absence of any significant magnetic contributions to $\Delta_{\text{mix}}G_m^{\text{excit}}$, the most important are the vibrational contributions.

We will use the Debye model, previously discussed in Chapter 3 for pure substances, in the high-temperature approximation. In this approximation, $\Delta_{\text{mix}}C_V = 0$ ($C_V = 3R$ for both alloy and elements) and $S_m^{\text{vib}} = -3R \log_e \theta_D$. Using this approximation, the following values for the vibrational contributions to the energy and entropy of mixing can be obtained:

$$\Delta_{\text{mix}}U_m^{\text{vib}} = \int \Delta_{\text{mix}}C_V \, dT = 0 \tag{16.5}$$

$$\Delta_{\text{mix}}S_m^{\text{vib}} = -3R \log_e \left(\frac{\theta_D(\text{A}_{1-x}\text{B}_x)}{\theta_D^{1-x}(\text{A})\theta_D^x(\text{B})} \right) = \text{const} \tag{16.6}$$

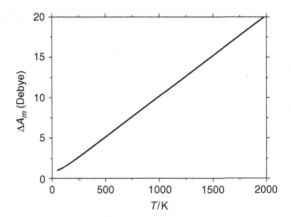

Figure 16.6 The $\Delta_{\text{mix}}A_m^{\text{vib}}$ for Debye model where $\theta_D = 200$ K for the pure metals and $\theta_D = 300$ K for alloy.

As an example, consider two metals, A and B, each with $\theta_D = 200$ K which form an alloy, $A_{1-x}B_x$, with $\theta_D = 300$ K. The magnitudes and differences of these Debye temperature values do not affect the conclusions. Figure 16.6 shows how $\Delta_{\text{mix}}A_m^{\text{vib}}$ varies with temperature. It can be seen that, in agreement with (16.5) and (16.6), there is a linear variation with temperature in the high-temperature region:

$$\Delta_{\text{mix}}G_m^{\text{excit}} \approx \Delta_{\text{mix}}A_m^{\text{vib}} = \text{const} \times T \qquad (16.7)$$

16.2.1   Coupling between Configurational and Thermal Excitations

Because of the vastly different time scales of the thermal excitational processes as compared with lattice configuration changes by atomic interchange, it is possible to *coarse grain* the alloy free energy by integrating over all the possible excitation states for just one configurational state. As a result, the configurational energy can be expressed as temperature-dependent effective cluster interactions; that is, for $\Delta_{\text{mix}}G_m$ we can write

$$\Delta_{\text{mix}}G_m = [\Delta_{\text{mix}}A_m^{\text{conf}} + \Delta_{\text{mix}}G_m^{\text{excit}}] + \Delta_{\text{mix}}G_m^{\text{asm}}$$
$$= [\Delta_{\text{mix}}U_m^{\text{conf}} + \Delta_{\text{mix}}A_m^{\text{vib}}] - T\Delta_{\text{mix}}S_m^{\text{conf}} + \Delta_{\text{mix}}G_m^{\text{asm}} \qquad (16.8)$$

where, from the Debye model in the high-temperature approximation at least, the square bracketed term is a linear function of temperature and is the one used in the configurational energy calculation.

Note, particularly, an unusual feature of (16.8). The square bracketed term for the temperature-dependent configurational energy includes an excitational free energy.

16.3   THE TOTAL GIBBS ENERGY IN EMPIRICAL MODEL CALCULATIONS

If it is assumed that the term from the atomic size mismatch contribution is parabolic and the configuration-dependent contribution is for a four-point cluster in the BW approximation, we can write the following expression for the total Gibbs energy of mixing:

$$\Delta_{\mathrm{mix}} G_m = \Omega x_A x_B + \sum_{\alpha\beta\gamma\delta} \sum_{ijkl} y_i^\alpha y_j^\beta y_k^\gamma y_l^\delta V_{ijkl}(T)$$

$$+ RT \sum_{(j)} \sum_i f^{(j)} y_i^{(j)} \log_e y_i^{(j)} \qquad (16.9)$$

Although (16.9) has been derived by considering ordered phases, it should also apply to the case where there is no long-range ordering and even when the cluster formation energies are positive. In this case, the mixing entropy term becomes that for a Raoultian ideal substitutional solution.

If we now compare the first two terms of this modeling equation with the Redlich–Kister polynomial expression previously used for these solutions,

$$G^{\mathrm{E}} = x_A x_B [L_0 + L_1(x_A - x_B) + L_2(x_A - x_B)^2 \cdots]$$

we can see that the regular solution parameter L_0 could arise from either the long-range configuration-independent term, $\Omega x_A x_B$, or the cluster energies, that is, from configuration-dependent terms. Given only the experimental thermodynamic properties of a phase, it is not possible to evaluate the individual contributions from these effects. In a more sophisticated treatment, both would be considered in a self-consistent way. The only reason for the separation into volume-independent, short-range chemical interactions and volume-dependent, atomic size mismatch effect has been to keep the model simple.

An example of the effect of introducing atomic size mismatch on the phase diagram is illustrated in Figure 16.7. In this example, for the A2/B2 ordering transformation in bcc alloys, both terms have been taken to be of similar magnitude. In the absence of the configuration-independent contribution, the order/disorder phase diagram appears as in Figure 16.7a. There is a second-order phase transition in this case. When a positive configuration-independent contribution is included, the phase diagram changes dramatically, as shown in Figure 16.7b. The second-order transition becomes a first-order transition at low temperatures. The junction point where the second order changes into a first order is called a *tricritical point*.

A real alloy which seems to exhibit this kind of behavior is found in the Fe–Ti system, the phase diagram for which is shown in Figure 16.8. Although the whole of the order/disorder phase diagram is not actually visible in this case because of the intrusion of other phases, it is possible to see similarities of the A2/B2 transition with that shown in Figure 16.7b.

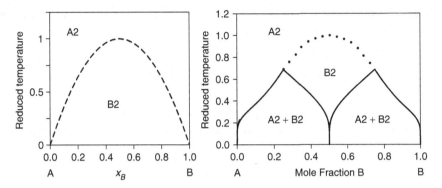

Figure 16.7 Calculated order/disorder phase diagrams, two-sublattice model, BW approximation: (*a*) without atomic size mismatch; (*b*) with atomic size mismatch.

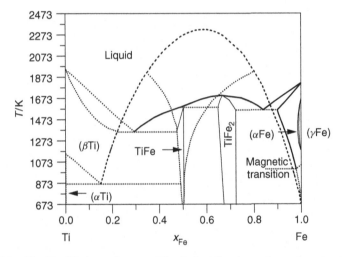

Figure 16.8 The Fe–Ti phase diagram. The dashed line is a schematic extension for the A2/B2 phase equilibrium.

EXERCISES

16.1 Given the following properties for Cr and Ni at room temperature, calculate an approximate parabolic relation of the form $\Delta_{\text{mix}}G_m^{\text{asm}} = \Omega x_A x_B$ for the effect of atomic size mismatch.

	V_m/cm^3 mol^{-1}	B_T/GPa
Cr	7.26	167
Ni	6.592	186.5

16.2 The following experimental results were obtained for the system (A,B) in which only one ordered phase is formed at midcomposition:

(a) $\Delta_{mix}H_m$ (random) $= -2.5$ kJ/mol alloy

(b) $\Delta_{mix}H_m$ (fully ordered) $= -7.5$ kJ/mol alloy

(c) $T_c = 1203$ K

If a configuration-independent Gibbs energy is represented by $\Delta_{mix}G_m$ (conf ind) $= \Omega x(1-x)$, what values of W_{AB} and Ω can reconcile the three sets of results?

17 Chemical Equilibria I: Single Chemical Reaction Equations

17.1 INTRODUCTION

In the next four chapters we will consider the thermodynamics of chemical equilibria. We begin by considering a single-chemical-reaction equation involving gaseous species only and introduce some different methods for solving the general problem of calculating the equilibrium state given the input amounts and the conditions (usually constant p and T).

It is important to note that this chapter is entitled single-chemical-reaction equations and not single chemical reactions. The distinction is very important in thermodynamics. In thermodynamics, when we write an equation like $CO(g) + 0.5O_2(g) = CO_2$, we are not suggesting that this chemical reaction is actually occurring in the system. This may or may not be true and, in any case, a concern for an actual reaction mechanisms is irrelevant in thermodynamics. The writing of this particular equation tells us that we have defined the system as consisting of three molecular species made from two types of elements and that we can hope to experimentally measure or calculate the amounts of these species at equilibrium under the imposed conditions. The chemical reaction equation is simply a statement of the stoichiometric relations between the molecular species defined as constituting the system. This distinction between a chemical reaction and a chemical reaction equation will become clearer when we consider a system comprised of several species, which requires the writing of more than one reaction equation.

17.2 THE EMPIRICAL EQUILIBRIUM CONSTANT

Consider the reaction equation

$$CH_4(g) + H_2O(g) = CO(g) + 3H_2(g) \tag{17.1}$$

Materials Thermodynamics. By Y. Austin Chang and W. Alan Oates
Copyright 2010 John Wiley & Sons, Inc.

The *empirical* equilibrium constant K_p for this reaction equation is an experimental quantity and is evaluated from

$$K_p = \left(\frac{p_{CO}^{eq} (p_{H_2}^{eq})^3}{p_{CH_4}^{eq} p_{H_2O}^{eq}} \right) \tag{17.2}$$

where the superscripts refer to the fact that we are talking about equilibrium conditions. Note that, when it is written in this way, K_p may or may not be dimensionless. It is not, for example, in the case of the particular example given in (17.2). We will have more to say on this point later.

Equation (17.2) can be generalized and written compactly as

$$\boxed{K_p = \prod_j p_j^{\nu_j, eq}} \quad \text{gaseous reaction equations} \tag{17.3}$$

where the stoichiometric coefficient ν_i is positive for product species and negative for reactant species.

It is also useful to define an empirical reaction equation quotient, Q_p, which applies to the reaction equation for any values of the partial pressures:

$$Q_p = \prod_j p_j^{\nu_j} \quad \text{gaseous reaction equations} \tag{17.4}$$

where Q_p is useful when discussing driving forces for reaction equations.

17.3   THE STANDARD EQUILIBRIUM CONSTANT

17.3.1   Relation to $\Delta_r G^{\circ}$

In the case of phase equilibrium, the condition of equilibrium between two phases α and β is given by

$$\mu_i^{\alpha} = \mu_i^{\beta} \quad \forall \, i \tag{17.5}$$

This equilibrium condition has to be modified when stoichiometry within the system is constrained by a chemical reaction equation. The general condition is readily appreciated by considering a specific example. If the system is defined to contain the species $SO_2(g)$, $O_2(g)$, and $SO_3(g)$, the reaction equation constraining these species is $SO_2(g) + 0.5O_2(g) = SO_3(g)$. As is shown later in this chapter [see (17.43)], the condition of equilibrium is now given by

$$\mu_{SO_3} - \mu_{SO_2} - \tfrac{1}{2}\mu_{O_2} = 0$$

The general condition for equilibrium when a chemical reaction equation constraint is present has to be modified to read

$$\boxed{\sum_i \nu_i \mu_i = 0} \quad \text{single-chemical-reaction equation} \qquad (17.6)$$

Although we have considered a single reaction equation in a single (gas) phase, the effect of reaction equation constraints is similar when the reaction equation involves more than one phase. The condition for phase equilibrium, (17.5), is seen to correspond with all the ν_i being ± 1.

In order to be able to express (17.6) in a more useful form, we can make use of the relation previously introduced for a component of a perfect gas mixture:

$$\mu_j = \mu_j^\circ + RT \log_e \left(\frac{p_j}{p_j^\circ} \right) \qquad (17.7)$$

Substituting this into (17.6), the equilibrium condition for the reaction can be expressed as:

$$\sum_j \nu_j \left[\mu_j^\circ + RT \log_e \left(\frac{p_j}{p_j^\circ} \right) \right] = 0 \quad \text{perfect gas mixtures only} \qquad (17.8)$$

or, equivalently,

$$\sum_j \nu_j \mu_j^\circ + RT \log_e \left[\prod_j \left(\frac{p_j}{p_j^\circ} \right)^{\nu_j} \right] = 0 \qquad (17.9)$$

Now the first term on the left-hand side is just the standard Gibbs energy of reaction, which refers to the complete conversion of reactants to products at 1 bar:

$$\sum_j \nu_j \mu_j^\circ = \Delta_r G^\circ \qquad (17.10)$$

so that (17.9) can be written as

$$\Delta_r G^\circ + RT \log_e \left[\prod_j \left(\frac{p_j}{p_j^\circ} \right)^{\nu_j} \right] = 0 \qquad (17.11)$$

If we now define a new quantity, the *standard* equilibrium constant K°, using

$$\Delta_r G^\circ = -RT \log_e K^\circ \qquad (17.12)$$

then we see, from (17.11),

$$-RT \log_e K^\circ + RT \log_e \left[\prod \left(\frac{p_j}{p_j^\circ} \right)^{\nu_j} \right] = 0 \quad \text{perfect gas mixtures only}$$

(17.13)

or

$$K^\circ = \exp \left(\frac{-\Delta_r G^\circ}{RT} \right) = \prod_j \left(\frac{p_j}{p_j^\circ} \right)^{\nu_j}$$

(17.14)

Note that K° is dimensionless, whereas the empirical equilibrium constant K_p, obtained from (17.3), may not be. If, however, the partial pressures are written as the ratio p_j/p_j° in K_p, then it, like K°, is also dimensionless so that the two quantities can be compared. In the special case of perfect gas mixtures only,

$$\boxed{K_p = K^\circ = \exp \left(\frac{-\Delta_r G^\circ}{RT} \right)}$$

(17.15)

that is, the *experimental* property K_p can be predicted from the independently *calculated* thermodynamic quantity K° under conditions which usually prevail in MS&E.

Equation (17.15) can also be used for the case of equilibrium between pure condensed phases and perfect gas mixtures. In this situation, the evaluation of K_p is restricted to the gas mixture partial pressures only since for the pure solid we can write, as previously, $\mu_i^* \approx \mu_i^\circ$ (the asterisk refers to the pure substance reference state at a pressure other than the standard 1 bar).

As an example, consider the reaction equation $2C(s) + O_2(g) = 2CO(g)$ for which we may write $K_p = p_{CO}^2/p_{O_2}$, involving the gas phase only, but $\Delta_r G^\circ$ used in calculating K° refers to that for the chemical reaction equation involving both gas and solid.

17.3.1.1 The Effect of Temperature on the Equilibrium Constant. The temperature variation of the standard equilibrium constant can be obtained by considering the temperature variation of $\Delta_r G^\circ / T$. Since R is a constant,

$$\frac{d(\Delta_r G^\circ / T)}{d(1/T)} = -T^2 \left(\frac{d(\Delta_r G^\circ / T)}{dT} \right)$$

(17.16)

$$= -T^2 \left[\frac{1}{T} \frac{d \Delta_r G^\circ}{dT} - \frac{\Delta_r G^\circ}{T^2} \right]$$

(17.17)

$$= T \Delta_r S^\circ + \Delta_r G^\circ$$

(17.18)

$$= \Delta_r H^\circ$$

(17.19)

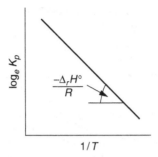

Figure 17.1 Standard enthalpy of reaction can be obtained from the slope of a plot of $\log_e K_p$ versus reciprocal temperature for reaction equations involving perfect gas mixtures and pure solids.

where we have used $\Delta_r G^\circ = \Delta_r H^\circ - T \Delta_r S^\circ$ and

$$\left(\frac{\partial \Delta_r G^\circ}{\partial T} \right)_p = -\Delta_r S^\circ$$

Then, from the relation between $\Delta_r G^\circ$ and K°, the above equation can be re-written as

$$\frac{d \log_e K^\circ}{d (1/T)} = -\frac{\Delta_r H^\circ}{R} \qquad \text{general relation} \qquad (17.20)$$

and, when it is safe to assume that $K_p = K^\circ$,

$$\boxed{\frac{d \log_e K_p}{d (1/T)} = -\frac{\Delta_r H^\circ}{R}} \qquad \text{perfect gas mixtures only} \qquad (17.21)$$

Equation (17.21) is known as the *Gibbs–Helmholtz* equation. As shown in Figure 17.1, $\log_e K_p$ for a chemical reaction equation is plotted as a function of $1/T$ over a small range of temperature, we expect to obtain a straight line in circumstances where the perfect gas assumption holds. In this case, the slope yields the negative of the standard enthalpy of reaction, valid for the temperature interval investigated. In practice, it is unlikely that $\Delta_r C_p^\circ$ will be exactly zero so that we should not expect to obtain a perfectly straight line over a large range of temperature, although, because of uncertainty in the data, we may go ahead and fit the results to a linear equation from which we can obtain an approximate value for $\Delta_r H^\circ (T)$.

17.3.2 Measurement of $\Delta_r G^\circ$

Previously, we have seen how $\Delta_r G^\circ$ can be obtained from a combination of substance and reaction calorimetric measurements alone.

But we can also make use of $\Delta_r G^\circ = -RT \log_e K^\circ$ if we work in a regime where it is safe to assume that $K_p = K^\circ$, that is, high temperatures and low pressures. This has been discussed briefly in Chapter 6. Here, we will give a little more detail.

We will illustrate with two examples, one involving phase equilibrium and the other chemical equilibrium:

1. Suppose we set out to measure the vapor pressure of Ag(l). Note that this is not straightforward experimentally. We cannot just connect the Ag at high temperature to a manometer at room temperature since this would result in the distillation of the Ag from the hot region to the cold region. Nevertheless, there are practical ways of overcoming this kind of problem. Knudsen effusion mass spectroscopy (KEMS) is another useful technique for measuring very low vapor pressures and can, in certain circumstances, lead to the simultaneous measurement of both of the component vapor pressures in a binary alloy, thereby giving a check with Gibbs–Duhem calculations.

   For the phase equilibrium represented by Ag(l) = Ag(g), we have $\Delta_{vap} G^\circ(\text{Ag}, T) = -RT \log_e(p_{\text{Ag}}/\text{bar})$ so that the vapor pressure measurement gives us the thermodynamic quantity $\Delta_{vap} G^\circ(\text{Ag}, T)$ and, if the equilibrium is measured as a function of T, $\Delta_{vap} H^\circ(\text{Ag}, T)$ and $\Delta_{vap} S^\circ(\text{Ag}, T)$ can be obtained.

2. The K_p for the reaction equation

$$\text{Fe(s)} + \text{H}_2\text{O(g)} = \text{FeO(s)} + \text{H}_2\text{(g)}$$

   can be measured in an apparatus similar to that in Figure 17.2. After evacuation, the Fe(s) sample is exposed to the water vapor from the $\text{H}_2\text{O(l)}$ reservoir. The formation of FeO(s) from Fe(s) takes place with the formation of $\text{H}_2\text{(g)}$. The controlled temperature of the $\text{H}_2\text{O(l)}$ fixes $p_{\text{H}_2\text{O}}$ throughout the apparatus. It is also the coldest part of the apparatus so that there is no chance of condensation elsewhere. The manometer reading gives $p_{\text{H}_2} + p_{\text{H}_2\text{O}}$ so that the individual p_{H_2} and $p_{\text{H}_2\text{O}}$ values can be obtained. Hence $K_p = p_{\text{H}_2}/p_{\text{H}_2\text{O}}$ is known. With the justified assumption that $K^\circ = K_p$, $\Delta_r G^\circ$ for the above reaction equation can be evaluated. If $\Delta_r G^\circ$ for the reaction equation

$$\text{Fe(s)} + \text{H}_2\text{O(g)} = \text{FeO(s)} + \text{H}_2\text{(g)}$$

   is combined with that for the reaction equation

$$\text{H}_2\text{(g)} + \tfrac{1}{2}\text{O}_2\text{(g)} = \text{H}_2\text{O(g)}$$

   for which $\Delta_r G^\circ$ is known with great accuracy, it then becomes possible to obtain $\Delta_f G^\circ(\text{FeO(s)}, T)$ from the equilibrium measurements. When $\Delta_r G^\circ$

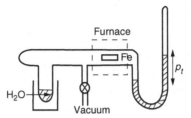

Figure 17.2 Simple apparatus for measurement of equilibrium for chemical reaction equation $Fe(s) + H_2O(g) = FeO(s) + H_2(g)$.

is measured as a function of T, both $\Delta_f H^\circ(FeO(s),T)$ and $\Delta_f S^\circ(FeO(s),T)$ can be obtained.

In most of the databases available for pure substances, approximately one-half of the data for $\Delta_f G^\circ$ have come from calorimetric measurements and the other half from equilibrium measurements similar to those described above.

17.4   CALCULATING THE EQUILIBRIUM POSITION

Calculating the final chemical equilibrium from a set of chosen initial amounts of substances and system conditions can be approached in exactly the same manner as has been described previously for phase equilibrium. The whole system, comprising either one or several phases and which is assumed to be closed and at constant total p and T, leads toward a minimization of the Gibbs energy in its approach to equilibrium. When chemical reaction equations are involved, however, the constraints on the minimization are different from those found for the case of phase equilibrium since the element mass balance for each element may involve more than one species containing that element and these may be in more than one phase.

A straightforward minimization of G subject to the constraints is not the only way of calculating the position of chemical equilibrium and, in the following, we will describe three different methods for calculating the equilibrium in a system which contains species related by a single reaction equation.

For illustration purposes, we will refer to the reaction equation

$$SO_2(g) + \tfrac{1}{2}O_2(g) = SO_3(g) \qquad (17.22)$$

Note that, since there is a change in the total amounts of substance in going from reactants to products in this particular reaction equation, the position of equilibrium depends on the total pressure.

(A) *Using the Mass Action Law*   For this example, the element mass balances for S and O and for the total number of species moles, n_{tot}, can be written as follows:

$$n_S = n^0_{SO_2} + n^0_{SO_3} = n_{SO_2} + n_{SO_3} \tag{17.23}$$

$$n_O = 2n^0_{SO_2} + 2n^0_{O_2} + 3n^0_{SO_3} = 2n_{SO_2} + 2n_{O_2} + 3n_{SO_3} \tag{17.24}$$

$$n_{tot} = n_{SO_2} + n_{O_2} + n_{SO_3} \tag{17.25}$$

where n^0_j represents the initial amount of species j and n_{tot} the total amounts of substances in the gas phase. The gaseous species partial pressures are related to the amounts and total pressure by

$$p_j = x_j p_{tot} = \left(\frac{n_j}{n_{tot}}\right) p_{tot} \tag{17.26}$$

The equilibrium condition (17.6) for this reaction is

$$\mu_{SO_3} - \mu_{SO_2} - \tfrac{1}{2}\mu_{O_2} = 0 \tag{17.27}$$

which, for perfect gas mixtures, can be rewritten as

$$\left[\mu^\circ_{SO_3} - \mu^\circ_{SO_2} - \tfrac{1}{2}\mu^\circ_{O_2}\right] + RT \log_e \left(\frac{p_{SO_3}}{p_{SO_2} \times p_{O_2}^{1/2}}\right) = 0 \tag{17.28}$$

or

$$\Delta_r G^\circ + RT \log_e \left(\frac{p_{SO_3}}{p_{SO_2} \times p_{O_2}^{1/2}}\right) = 0 \tag{17.29}$$

or

$$K^\circ = \left(\frac{p_{SO_3}}{p_{SO_2} \times p_{O_2}^{1/2}}\right) \tag{17.30}$$

This last equation is usually known as the mass action law when K° is replaced wth K_p.

The mass balance requirements (17.23)–(17.25), the partial pressure conversion (17.26), and the equilibrium condition (17.30) may be used to calculate the individual equilibrium partial pressures. The solution requires solving several nonlinear simultaneous equations.

In this simple case of a single reaction, a hand calculation may be carried out by using the mass balance equations to eliminate two of the species, for example,

$$n_{SO_2} = n_S - n_{SO_3}$$

$$n_{O_2} = \tfrac{1}{2}(n_O - 2n_S - n_{SO_3})$$

and substitute the resulting partial pressures from (17.26) into (17.30) to solve for p_{SO_3}. The other equilibrium partial pressures can then be obtained subsequently.

(B) *Using the Extent of Reaction, ξ*   In this method the element mass balance equations are circumvented by taking advantage of the fact that the changes in the species amounts, dn_j, during a reaction are not independent but are related through the stoichiometry of the reaction equation. As a result, only one parameter per reaction equation is required to specify the changes in the amounts of the species. The differential changes in the species amounts for the reaction equation under consideration are seen to be

$$\frac{dn_{SO_2}}{-1} = \frac{dn_{O_2}}{-1/2} = \frac{dn_{SO_3}}{+1} = d\xi \tag{17.31}$$

The variable ξ is called the *extent of reaction* (units are moles), and it can be seen that this one variable expresses the changes in the amount of any species, j, in the chemical reaction equation, that is, $dn_j = v_j\, d\xi$, which on integration gives

$$n_j = n_j^0 + v_j\xi \tag{17.32}$$

where ξ is taken to be zero in the initial state. For the reaction equation (17.22) the input amounts are given by:

$$n_{SO_2} = n_{SO_2}^0 - \xi \tag{17.33}$$

$$n_{O_2} = n_{O_2}^0 - \tfrac{1}{2}\xi \tag{17.34}$$

$$n_{SO_3} = n_{SO_3}^0 + \xi \tag{17.35}$$

$$n_{tot} = n_{SO_2}^0 + n_{O_2}^0 + n_{SO_3}^0 - \tfrac{1}{2}\xi \tag{17.36}$$

Equations (17.33)–(17.36) together with (17.26) and (17.30) may be solved simultaneously to obtain the individual partial pressures. Alternatively, the equilibrium constant can be expressed solely in terms of ξ and the starting amounts of species.

(C) *Using Lagrangian Multipliers*   With both of the previous methods it is necessary to know the chemical reaction equation. While this causes no problem in a simple system, the advantage of the method based on Lagrangian multipliers is that a reaction equation is not required. It is only necessary to specify the species designated as comprising the system. The element mass balance equations, for example, (17.23)–(17.25), are also used once the system is specified.

The introduction of the Lagrangian multipliers λ_i transforms the constrained minimization of the $G = \sum_j n_i\mu_j$ problem into an unconstrained minimization of a Lagrangian function $\mathcal{L}$ with Lagrangian multipliers assigned to the element mass balances.

In this example,

$$\lambda_S: \qquad n_S - (n_{SO_2} + n_{SO_3}) = 0 \tag{17.37}$$

$$\lambda_O: \qquad n_O - (2n_{SO_2} + 2n_{O_2} + 3n_{SO_3}) = 0 \tag{17.38}$$

and the Lagrangian is

$$\mathcal{L} = \sum_j n_j \mu_j + \lambda_S[n_S - (n_{SO_2} + n_{SO_3})]$$

$$+ \lambda_O[n_O - (2n_{SO_2} + 2n_{O_2} + 3n_{SO_3})] \tag{17.39}$$

Minimization of $\mathcal{L}$ with respect to the amounts of the molar species gives the following equations for the minimum:

$$\frac{\partial \mathcal{L}}{\partial n_{SO_2}} = \mu_{SO_2} - \lambda_S - 2\lambda_O = 0 \tag{17.40}$$

$$\frac{\partial \mathcal{L}}{\partial n_{O_2}} = \mu_{O_2} - 2\lambda_O = 0 \tag{17.41}$$

$$\frac{\partial \mathcal{L}}{\partial n_{SO_3}} = \mu_{SO_3} - \lambda_S - 3\lambda_O = 0 \tag{17.42}$$

These equations, in conjunction with the mass balances (17.23)–(17.25), the partial pressure equation (17.26), and the chemical potential expression (17.7), may be solved simultaneously to obtain the equilibrium partial pressures.

Although they are not required in the evaluation of the partial pressures, elimination of the Lagrangian multipliers from (17.40)–(17.42) gives the relation between the chemical potentials which hold at equilibrium when the Gibbs energy is at its minimum. This relation is, of course,

$$\mu_{SO_3} - \mu_{SO_2} - \tfrac{1}{2}\mu_{O_2} = 0 \tag{17.43}$$

Naturally, all three methods described give the same answer and all could, in principle, be used for the more complex situations where more than one reaction equation must be introduced. The use of Lagrangian multipliers, however, has the clear advantage of not requiring a set of independent reaction equations to be specified beforehand and is, therefore, the method usually preferred in computer calculations on more complex systems.

Because of its importance, we will generalize the approach described above for a specific example by introducing the *conservation (or formula) matrix*, **A**, which lists the elements (number M) as columns and the species (number N) as rows. The matrix stores the mass balances.

For the above example system the conservation matrix is written as

$$
\mathbf{A} = \begin{pmatrix}
 & SO_2 & O_2 & SO_3 \\
S & 1 & 0 & 1 \\
O & 2 & 2 & 3
\end{pmatrix}
$$

where the member A_{ij} gives the number of i elements in the jth species.

In general, the mass balance relations can be written in terms of the members of $\mathbf{A}$ as

$$
n_i - \sum_{j=1}^{N} A_{ij} n_j = 0 \quad \forall \text{ elements } i \tag{17.44}
$$

and the Lagrangian function as

$$
\mathcal{L} = \sum_{j=1}^{N} n_j \mu_j + \sum_{i=1}^{M} \lambda_i \left(n_i - \sum_{j=1}^{N} A_{ij} n_j \right) \tag{17.45}
$$

where M is the number of elements and N the number of species. This equation is a generalization of (17.39).

Differentiation of the Lagrangian with respect to the amount of each species and setting the derivatives to zero gives the equilibrium conditions

$$
\frac{\partial \mathcal{L}}{\partial n_j} = \mu_j + \sum_{i}^{M} \lambda_i A_{ij} = 0 \quad \forall \text{ elements } i \tag{17.46}
$$

which is a generalization of (17.40)–(17.42).

17.5   APPLICATION OF THE PHASE RULE

We have seen that, when chemical reaction equations have to be considered, the constraints on the chemical potentials are different from those met when considering phase equilibria. This influences the number of *independent components*, C, to be used in the phase rule, $F = C - \phi + 2$. In the case of alloy-phase equilibria, we simply equated C with the number of elements. The number of constraints on the $\mu_i^\alpha = \mu_i^\beta$ relations is $C(\phi - 1)$ (see Chapter 8). There is also a Gibbs–Duhem constraint for each phase.

In the case of a single-chemical-reaction equation involving N species there is just the one constraint given in (17.6), namely, $\sum_{j}^{N} v_j \mu_j = 0$, so that the number of independent components $C = N - 1$. The phase rule for this case is then modified to read

$$
F = (N - 1) - \phi + 2 = N - \phi + 1 \tag{17.47}
$$

If we apply this to the example of the homogeneous gas reaction equation

$$SO_2(g) + \tfrac{1}{2}O_2(g) = SO_3(g)$$

then

$$F = 3 - 1 + 1 = 3 \tag{17.48}$$

and we see that, if we fix p_{tot}, T, and the initial amount of one substance, the system is fully specified.

In the case of the reaction equation $CaCO_3(s) = CaO(s) + CO_2(g)$, we have to consider a three-phase system so that

$$F = 3 - 3 + 1 = 1 \tag{17.49}$$

and fixing T is sufficient to specify the equilibrium state.

We will return to the application of the phase rule to more complex chemical reaction equation situations in the next chapter.

EXERCISES

17.1 A gas mixture of 50% CO, 25% CO_2, and 25% H_2 (percentages by volume) is fed into a furnace at 900°C. Find the composition of the equilibrium $CO–CO_2–H_2O–H_2$ gas if the total pressure in the furnace is 1 bar.

17.2 How much heat is evolved when 1 mol of SO_2 and $\tfrac{1}{2}$ mol of O_2 react to form the equilibrium $SO_3–SO_2–O_2$ mixture at 1000 K and at a total pressure of 1 bar?

17.3 What is the oxygen partial pressure exerted by an equilibrium gas mixture of $CO_2–CO–H_2–H_2O$ produced at 1600°C by mixing CO_2 and H_2 in the ratio of 3 : 1?

17.4 If you wish to make a mixture of $CO_2–CO–H_2–H_2O$ at 1600°C with an oxygen partial pressure of 10^{-7} bar at a total pressure of 1 bar, what is the initial ratio of CO_2 and H_2 needed to produce this equilibrium mixture?

17.5 (a) Calculate $\Delta_r G^\circ(1000\ \text{K})$ for the reaction

$$CH_4(g) + CO_2(g) = 2CO(g) + 2H_2(g)$$

(b) Assuming that $K^\circ = K_p$, at what temperature does $K_p = 1$?

(c) In which direction does the equilibrium shift when:
  (i) The temperature of an equilibrated $CH_4–CO_2–CO–H_2$ gas mixture is increased?
  (ii) The total pressure is decreased?

17.6 By establishing the equilibrium $PCl_5 = PCl_3 + Cl_2$ in a mixture of PCl_5 and PCl_3, it is required to obtain a partial pressure of Cl_2 of 0.1 bar at 500 K when the total pressure is 1 bar. In what ratio must PCl_5 and PCl_3 be mixed?

17.7 A producer of copper parts uses the powder metallurgy technique to make these parts. To increase the strength of these parts, they are dispersion hardened with fine Al_2O_3 powders.

  The producer can buy Cu powder containing 2% Fe much cheaper than the pure Cu powder he has been using. (The less pure Cu powder is a by-product of the cement copper process.) Also, he finds that magnetite (Fe_3O_4) is almost as effective in dispersion hardening of copper as Al_2O_3.

  If he sinters the less pure powder in the normal H_2 atmosphere (or vacuum), the result is a (Cu, Fe) alloy containing 2% Fe. Such an alloy is unsuitable for the requirements of his parts, however. Describe an oxidizing/sintering step that he can use in order to internally oxidize the Fe to produce a relatively pure Cu part in which are well dispersed fine iron oxide particles. What is the lowest a_{Fe} obtainable in the Cu solid solution? Just how, in practice, would be a reasonable way to conduct the internal oxidation step?

  Assume (Cu, Fe) solid solution behaves like a strictly regular solution with $L_0 \approx 30,750$ J mol^{-1}.

18 Chemical Equilibria II: Complex Gas Equilibria

Most chemically reacting systems of interest in MS&E are complex so that it is rare that such systems can be expressed in terms of just the one chemical reaction equation, as was discussed in Chapter 17. Atmospheres generated from fossil fuel combustion, for example, contain, principally, molecules, atoms, and ions containing the elements C, H, O, S, and N and the number of possible molecular species based on these five elements can be extremely large. The most familiar are H_2, O_2, S_2, N_2, H_2O, H_2S, NH_3, CH_4, CO, CO_2, SO_2, and SO_3 but there are countless others, often including undesirable dioxins as well as elements like U, Th, and As.

Figure 18.1 shows the gas phase composition resulting from the combustion of methane in oxygen as function of temperature. Considerably more complex gas phase compositions can result, for example, in Chemical Vapor Deposition processes.''

System complexity is also the norm in the metal-processing industries. There, it is common to find more than two phases in contact, for example, in the copper extraction industry the system may comprise gas, slag, matte, and metal phases, each of which is chemically complex, with material passing between the phases by chemical interaction in the attempt to reach thermodynamic equilibrium.

The keyword to be borne in mind here is complexity. We need to learn how thermodynamics can be gainfully applied to such complex, chemically reacting systems. As a preliminary, in this chapter, we consider the calculation of equilibrium in complex gas mixtures. Chemical equilibria involving gases and condensed phases will be discussed in Chapters 19 and 20.

18.1 THE IMPORTANCE OF SYSTEM DEFINITION

Any discussion of equilibrium in chemically reacting systems leads to an appreciation of what thermodynamics can and cannot do:

> Although thermodynamics is used to accurately predict properties or data from other properties or data, it can only make these predictions for the system defined by the thermodynamicist.

Materials Thermodynamics. By Y. Austin Chang and W. Alan Oates
Copyright 2010 John Wiley & Sons, Inc.

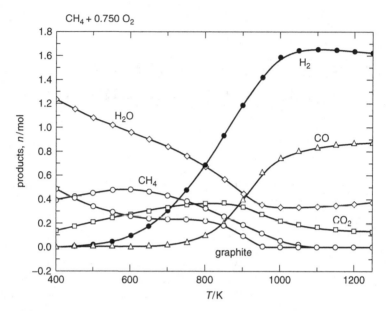

Figure 18.1   Calculated gas phase composition in the combustion of 1 mole of methane with 0.75 mole oxygen. Reproduced with permission from H. Yokokawa, *J. Phase Equilib.* **26**, p. 155, 2002.

A poorly defined system will lead to an incorrect prediction of properties. The most common way in which this is encountered is the failure to include some species in the system specification. This neglect, which may be due to ignorance or oversight, cannot be attributed to a failure of thermodynamics.

Problems associated with system definition are much less frequently encountered in the case of phase equilibrium calculations. Thus, if a calculated stable Al–Zn phase diagram were to be presented as shown in Figure 18.2*a*, one would quickly be alerted to the fact that the hcp phase had been overlooked in the calculation. When this oversight is rectified, the more familiar phase diagram shown in Figure 18.2*b* can then be calculated.

This type of neglect, however, with the user often being unaware of it, happens frequently in calculations of chemically reacting systems, particularly for those involving high-temperature calculations of heterogeneous equilibria.

Example 18.1 Combustion of $H_2(g)$ with $O_2(g)$

A very simple illustration of the importance of the role of the thermodynamicist in performing meaningful thermodynamic calculations arises in the calculation of the adiabatic flame temperature for the stoichiometric combustion of $H_2(g)$ with $O_2(g)$ at a total pressure of 1 bar. This calculation entails assuming that the reaction between $H_2(g)$ and $O_2(g)$ takes place in a closed system whose volume can vary in order to maintain a constant pressure of 1 bar. The combustion takes place without any heat flow to or from the surroundings. This results in the

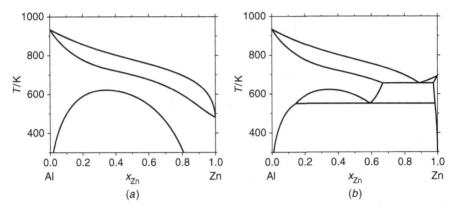

Figure 18.2 Calculated Al–Zn phase diagrams: (a) Without hcp phase taken into account; (b) with all phases taken into account.

enthalpy of the input gases being equal to the enthalpy of the product gases, that is, $\Delta_r H^\circ$ for the following reaction equation is zero:

$$H_2(g)(298.15 \text{ K}) + \tfrac{1}{2}O_2(g)(298.15 \text{ K}) = H_2O(g)(T) \qquad (18.1)$$

Consider the different outcomes from three different decisions by the thermodynamicist concerning the calculation:

(i) Thermodynamicist A assumes that the reaction goes to completion, that is, T is calculated from the enthalpy changes for the two following reaction equations:

$$H_2(g)(298.15 \text{ K}) + \tfrac{1}{2}O_2(g)(298.15 \text{ K}) = H_2O(g)(298.15 \text{ K})$$

$$\Delta H = \Delta_f H^\circ(298.15 \text{ K})$$

$$H_2O(g)(298.15 \text{ K}) = H_2O(g)(T) \qquad \Delta H = \int_{298.15 \text{ K}}^{T} C_p^\circ(H_2O)\, dT$$

(ii) Thermodynamicist B defines the system as comprising $H_2(g)$, $O_2(g)$, and $H_2O(g)$ but considers the possibility that the chemical reaction equation given in (18.1) does not go to completion, that is, all three species may be present at the final temperature. The relative amounts of the three gases are governed by $\Delta_r G^\circ$ at the final temperature as well as by mass balance constraints.

(iii) Thermodynamicist C defines the system as comprising all the gaseous species contained in a comprehensive database containing the elements H and O in addition to the species $H_2(g)$, $O_2(g)$, and $H_2O(g)$. All these extra species may be present at the final temperature.

The calculated flame temperature and partial pressure results obtained from these three different assumptions are as follows:

Species	p (bars)	Species	p (bars)
Flame Temperature:	**4894 K**	Flame Temperature:	**3146.31 K**
H_2O	1.0	H_2O	0.618
H_2	0.0	H_2	0.157
O_2	0.0	H	0.0964
		O_2	0.0774
Flame Temperature:	**3508.91 K**	O	0.0508
H_2O	0.569	H_2O_2	3.57×10^{-6}
H_2	0.287	O_3	4.19×10^{-8}
O_2	0.144		

It can be seen that the results obtained for the calculated adiabatic flame temperature differ markedly (the result obtained by thermodynamicist C will be close to the experimental value). The important point to appreciate is that the differences in the calculated results stem from the different decisions made by the thermodynamicists. They do not represent any failure on the part of thermodynamics.

This problem of system definition is exacerbated when more complex situations than this simple example are considered. We will encounter it again when we come to discuss equilibrium involving both gaseous and condensed phases. For the present, we emphasize that the failure to properly define the system in thermodynamic calculations of chemical equilibria is a common occurrence.

18.2   CALCULATION OF CHEMICAL EQUILIBRIUM

In Chapter 17 we examined some different approaches to the calculation of the position of chemical equilibrium for single-reaction equations involving perfect gas mixtures and pure solids. All these methods can be applied to more complex situations, but when the gases are not perfect and the condensed phases are nonideal solutions, then, just as in the case of phase equilibria, local minima in the Gibbs energy can arise. These local minima make the equilibrium calculations considerably more difficult. In this chapter we will restrict ourselves to some fairly straightforward examples which avoid these complications by considering the case of several species present in a perfect gas mixture. Even here, some calculation methods are preferable to others.

Specifically, we will illustrate with a system defined as containing the gaseous molecular species $CH_4(g)$, $O_2(g)$, $CO_2(g)$, $H_2O(g)$, $CO(g)$, and $H_2(g)$. Note that, although there are many other molecular species which contain the elements C, H, and O, we have elected to define the system as containing just these six species. We might elect to do this, for example, if we know that under the conditions where the results from the calculations are to be used it is known that the concentration of, say, $C_2H_2(g)$, is unimportant.

18.2.1   Using the Extent of Reaction

A principal objective of chemical reaction stoichiometry is to determine the number of independent chemical reaction equations, R, required to determine the equilibrium in the defined system. In this relatively simple case, it is possible to see, by inspection, that one set of independent chemical reaction equations is the following:

$$CH_4(g) + 2O_2(g) = CO_2(g) + 2H_2O(g) \tag{18.2}$$

$$2CO_2(g) = 2CO(g) + O_2(g) \tag{18.3}$$

$$CH_4(g) + O_2(g) = CO_2(g) + 2H_2(g) \tag{18.4}$$

so that, in this case, $R = 3$. This selection is, of course, not the only set of independent reaction equations which can be written involving these six species. The number of independent equations will, however, always be three.

When the number of species is much larger than that in our example, it is necessary to have a more general method for determining R. In order to do this, we first set up the *formula or conservation matrix* $\mathbf{A}$. For our example, this is an $M \times N$ matrix:

$$\mathbf{A} = \begin{pmatrix} & CH_4 & O_2 & CO_2 & H_2O & CO & H_2 \\ C & 1 & 0 & 1 & 0 & 1 & 0 \\ H & 4 & 0 & 0 & 2 & 0 & 2 \\ O & 0 & 2 & 2 & 1 & 1 & 0 \end{pmatrix}$$

The number of independent components, C, is given by rank($\mathbf{A}$). In this example, $C = 3$. The value of R can then be obtained from $R = N - C$. In this example, $R = 6 - 3 = 3$.

Having obtained the number of independent reaction equations, it is necessary to assign an extent of reaction, ξ, to each of the chosen reaction equations. In our example, ξ_1, ξ_2, and ξ_3 are assigned to (18.2)–(18.4), respectively. Then we may proceed exactly as we did for a single reaction in Chapter 17. For example, for (18.2), we may write

$$\frac{dn_{CH_4}}{-1} = \frac{dn_{O_2}}{-2} = \frac{dn_{CO_2}}{+1} = \frac{dn_{H_2O}}{+2} = d\xi_1 \tag{18.5}$$

From this and similar equations for the reaction equations (18.3) and (18.4), we may integrate and obtain the values of the species amounts in terms of the three reaction variables:

$$n_{CH_4} = n_{CH_4}^o - \xi_1 - \xi_3 \tag{18.6}$$

$$n_{O_2} = n_{O_2}^o - 2\xi_1 + \xi_2 - \xi_3 \tag{18.7}$$

$$n_{CO_2} = n_{CO_2}^o + \xi_1 - 2\xi_2 + \xi_3 \tag{18.8}$$

$$n_{H_2O} = n_{H_2O}^o + 2\xi_1 \tag{18.9}$$

$$n_{CO} = n_{CO}^o + 2\xi_2 \tag{18.10}$$

$$n_{H_2} = n_{H_2}^o + 2\xi_3 \tag{18.11}$$

$$n_{tot} = n_{CH_4}^o + n_{O_2}^o + n_{CO_2}^o + n_{H_2O}^o + n_{CO}^o + n_{H_2}^o + \xi_2 + \xi_3 \tag{18.12}$$

The partial pressures are related to the amounts and total pressure by

$$p_j = \left(\frac{n_j}{n_{tot}} \right) p_{tot} \tag{18.13}$$

and the equilibrium conditions for these reactions are

$$\Delta_r G_1^\circ + RT \log_e \left(\frac{p_{CO_2} \times p_{H_2O}^2}{p_{CH_4} \times p_{O_2}^2} \right) = 0 \tag{18.14}$$

$$\Delta_r G_2^\circ + RT \log_e \left(\frac{p_{CO}^2 \times p_{O_2}}{p_{CO_2}^2} \right) = 0 \tag{18.15}$$

$$\Delta_r G_3^\circ + RT \log_e \left(\frac{p_{CO_2} \times p_{H_2}^2}{p_{CH_4} \times p_{O_2}} \right) = 0 \tag{18.16}$$

The set of mass balance equations 18.6–18.12 plus the relation between molar amounts and the partial and total pressures, (18.13), together with the equilibrium relations (18.14)–(18.16) can be used to solve for the final chemical equilibrium.

Figure 18.3 shows, qualitatively, how the Gibbs energy varies as a function of ξ_1 and ξ_2 for a system defined by two independent reaction equations. The equilibrium position is given by the minimum in the Gibbs energy, which can be obtained by satisfying

$$\left(\frac{\partial G}{\partial \xi_1} \right) = \left(\frac{\partial G}{\partial \xi_2} \right) = 0 \tag{18.17}$$

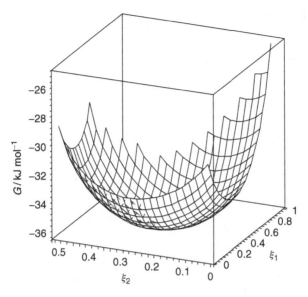

Figure 18.3 Variation of G as function of ξ_1 and ξ_2, drawn using Maple®.

18.2.2 Using Lagrangian Multipliers

In the example which involves only six species and three independent reactions, application of the extent of the reaction method is straightforward. When the system comprises on the order of 100 molecular species derived from just a few elements, it is more difficult and the method based on Lagrangian multipliers is to be preferred.

A Lagrangian multiplier, λ_i, is assigned to the mass balance for each element:

$$\lambda_C : \quad n_C - (n_{CO} + n_{CO_2} + n_{CH_4}) \tag{18.18}$$

$$\lambda_H : \quad n_H - (2n_{H_2} + 2n_{H_2O} + 4n_{CH_4}) \tag{18.19}$$

$$\lambda_O : \quad n_O - (n_{CO} + 2n_{O_2} + 2n_{CO_2} + n_{H_2O}) \tag{18.20}$$

and the Lagrangian function is written as

$$\mathcal{L} = \sum_{j=1}^{N} n_j \mu_j + \sum_{i=1}^{M} \lambda_i \left(n_i - \sum_{j=1}^{N} A_{ij} n_j \right) \tag{18.21}$$

Differentiation of this function with respect to the amount of each species and setting the derivatives to zero gives the equilibrium conditions. For our example, these partial derivatives are

$$\frac{\partial \mathcal{L}}{\partial n_{H_2}} = \mu_{H_2} - 2\lambda_H = 0 \tag{18.22}$$

$$\frac{\partial \mathcal{L}}{\partial n_{O_2}} = \mu_{O_2} - 2\lambda_O = 0 \tag{18.23}$$

$$\frac{\partial \mathcal{L}}{\partial n_{H_2O}} = \mu_{H_2O} - 2\lambda_H - \lambda_O = 0 \tag{18.24}$$

$$\frac{\partial \mathcal{L}}{\partial n_{CO}} = \mu_{CO} - \lambda_C - \lambda_O = 0 \tag{18.25}$$

$$\frac{\partial \mathcal{L}}{\partial n_{CO_2}} = \mu_{CO_2} - \lambda_C \quad 2\lambda_O = 0 \tag{18.26}$$

$$\frac{\partial \mathcal{L}}{\partial n_{CH_4}} = \mu_{CH_4} - \lambda_C - 4\lambda_H = 0 \tag{18.27}$$

The chemical potential for a component of a perfect gas mixture is given by

$$\mu_j = \mu_j^\circ + RT \, \log_e \left(\frac{p_j}{p_j^\circ} \right) \tag{18.28}$$

By using this equation, the element mass balances, $p_j = x_j p_{tot}$, and (18.22)–(18.27), we can obtain all the partial pressures.

Although it is not required in the equilibrium calculation, it can be seen that the elimination of the Lagrangian multipliers in (18.22)–(18.27) gives three independent relations between the chemical potentials which hold at equilibrium. One possible set is

$$\mu_{CO_2} + 2\mu_{H_2O} - \mu_{CH_4} - 2\mu_{O_2} = 0 \tag{18.29}$$

$$2\mu_{CO} + \mu_{O_2} - 2\mu_{CO_2} = 0 \tag{18.30}$$

$$\mu_{CO_2} + 2\mu_{H_2O} - \mu_{CH_4} - 2\mu_{O_2} = 0 \tag{18.31}$$

Equations (18.29)–(18.31) can be interpreted as referring to the equilibrium conditions for the reaction equations already given in (18.2)–(18.4).

But the three relations (18.29)–(18.31) are not the only three possible independent relations which can be obtained from (18.22)–(18.27) by elimination of the Lagrangian multipliers. For example, we could just as easily have obtained

$$2\mu_{H_2} + 2\mu_{CO} - \mu_{CH_4} - \mu_{CO_2} = 0$$

$$\mu_{CO} + 3\mu_{H_2} - \mu_{CH_4} - \mu_{H_2O} = 0$$

$$2\mu_{H_2O} - 2\mu_{H_2} - \mu_{O_2} = 0$$

This emphasizes the point that, although there are only three independent relations between the chemical potentials of the species, the relations can be selected in different ways. It also follows that there is nothing in a balanced chemical reaction equation which relates to any actual mechanisms of chemical reactions occurring within the system.

18.3   EVALUATION OF ELEMENTAL CHEMICAL POTENTIALS IN COMPLEX GAS MIXTURES

Usually, it is the behavior of one or, occasionally, two particular elements in a complex gas atmosphere which is of primary interest. We may wish to know, for example, whether a gas at a given p, T and initial composition is oxidizing, reducing, sulfurizing, carburizing, nitriding, and so on, to the particular condensed phase the gas phase is contacting. Complete knowledge of the chemical analysis of the gas becomes of secondary importance in these situations.

If we concentrate on the oxidizing/reducing power of a gas mixture for the moment, then it is the chemical potential of oxygen which is of interest. At high oxygen potentials the gas is likely to be oxidizing, at low oxygen potentials reducing. The oxygen potential is defined (in a perfect gas mixture) by

$$\Delta\mu_{O_2} = \mu_{O_2} - \mu_{O_2}^{\circ} = RT \, \log_e \left(\frac{p_{O_2}}{p_{O_2}^{\circ}} \right) \tag{18.32}$$

It is often the case that the partial pressure of $O_2(g)$ is much too small to be obtained from chemical analysis. In such cases it is useful to be able to get a handle on its magnitude by other means.

No physical meaning was given to the Lagrangian parameters in (18.22)–(18.27) since this was unnecessary in the solution of the chemical equilibrium problem. But it is clear from these equations that the Lagrangian parameters are in fact chemical potentials of the elements. For example, we may appreciate their meaning by considering (18.25):

$$\mu_{CO} - \lambda_C - \lambda_O = 0$$

which is equivalent to writing

$$\mu_{CO} - \mu_C - \mu_O = 0 \tag{18.33}$$

or

$$\mu_{CO} - \mu_C - \tfrac{1}{2}\mu_{O_2} = 0 \tag{18.34}$$

so that we already know the element chemical potentials when the Lagrangian method is used for calculating chemical equilibrium. They are just the Lagrangian parameters associated with the mass balances.

The relation of element chemical potentials to molecular species chemical potentials can also be seen directly by considering particular reaction equations. By doing this we can see how to obtain an element chemical potential from the molecular gas analysis. Consider, for example, the reaction equation

$$2CO(g) + O_2(g) = 2CO_2(g)$$

By using the relation between the standard Gibbs energy of reaction and the standard equilibrium constant and assuming that the latter is the same as the empirical equilibrium constant (perfect gas mixtures), we can write

$$\Delta_r G^\circ(T) = -2RT \, \log_e \left(\frac{p_{CO_2}}{p_{CO}} \right) + RT \, \log_e p_{O_2} \qquad (18.35)$$

Rearranging gives

$$\Delta \mu_{O_2}(T) = 2 \left[\Delta_f G^\circ(CO_2(g), T) - \Delta_f G^\circ(CO(g), T) \right]$$
$$+ 2RT \, \log_e \left(\frac{p_{CO_2}}{p_{CO}} \right) \qquad (18.36)$$

Similarly, by considering the reaction equation,

$$2H_2(g) + O_2(g) = 2H_2O(g)$$

we have

$$\Delta_r G^\circ(T) = -2RT \, \log_e \left(\frac{p_{H_2O}}{p_{H_2}} \right) + RT \, \log_e p_{O_2} \qquad (18.37)$$

$$\Delta \mu_{O_2}(T) = 2 \, \Delta_f G^\circ(H_2O(g), T) + 2RT \, \log_e \left(\frac{p_{H_2O}}{p_{H_2}} \right) \qquad (18.38)$$

This type of procedure can be used for other elements to specify the elemental chemical potentials in terms of the ratios of different constituents of a complex mixture.

For example, carbon potentials can be obtained from consideration of one of the following reaction equations:

$$C(s) + 2H_2(g) = CH_4(g) \qquad (18.39)$$
$$C(s) + CO_2(g) = 2CO(g) \qquad (18.40)$$

sulfur potentials from

$$S_2(g) + 2H_2(g) = 2H_2S(g) \qquad (18.41)$$
$$S_2(g) + 4SO_3(g) = 6SO_2(g) \qquad (18.42)$$

and nitrogen potentials from

$$N_2(g) + 3H_2(g) = 2NH_3(g) \tag{18.43}$$

18.4 APPLICATION OF THE PHASE RULE

If we apply the most frequently used form of the phase rule, (8.26), namely, $F = C - \phi + 2$, to the single-reaction equation $CO(g) + \frac{1}{2}O_2(g) = CO_2(g)$ and take the number of components to be the same as the number of molecular species, that is, three, then application of this equation gives $F = 3 - 1 + 2 = 4$, which is incorrect. If, on the other hand, we take $C = N - R = 3 - 1 = 2$, then $F = 2 - 1 + 2 = 3$, which is correct. If, for example, we fix p, T, and p_{CO_2}, then the two remaining partial pressures can be obtained from the equilibrium constant and the fact that the sum of the partial pressures is equal to the fixed total pressure.

Similar reasoning applies to more complex situations. In our example of a system comprising six molecular species, if we use $F = 6 - 1 + 2 = 7$, we obtain an incorrect answer. If, however, we use $C = N - R = 6 - 3 = 3$, then $F = 3 - 1 + 2 = 4$, which is correct. Fixing p, T, p_{O_2}, and p_{CO_2} leaves us with four unknowns. These can be obtained from the three equilibrium constants and the partial pressure sum.

The selection of *constituents* (the *independent* species) required to describe a system is not unique but their number is. The *number of components* is the same as the *number of constituents*, but the components may or may not involve the constituents.

The selected components may be these constituents but other selections may be made. The only requirement of a *component* is that it can be used to express composition over the range of interest for the system under consideration.

With some selections of components, it is possible for some compositions to have negative amounts of the selected components. As an illustration, consider the following selections of components for the Fe–O system. In the first we choose Fe and O as the components and in the second we choose FeO and Fe_2O_3.

The conservation matrices for these two choices are as follows:

$$\mathbf{A} = \begin{pmatrix} \dfrac{\text{phase}}{\text{component}} & Fe(s) & FeO(s) & Fe_3O_4(s) & Fe_2O_3(s) & O_2(g) \\ Fe & 1 & 1 & 3 & 2 & 0 \\ O & 0 & 1 & 4 & 3 & 2 \end{pmatrix}$$

$$\mathbf{A} = \begin{pmatrix} \dfrac{\text{phase}}{\text{component}} & Fe(s) & FeO(s) & Fe_3O_4(s) & Fe_2O_3(s) & O_2(g) \\ FeO & 3 & 1 & 1 & 0 & -4 \\ Fe_2O_3 & -1 & 0 & 1 & 1 & 2 \end{pmatrix}$$

Although the selection of components is different in these two formula matrices, their rank, and hence the number of independent components, remains the same.

EXERCISES

18.1 Acetylene, $C_2H_2(g)$, is burnt at room temperature with $O_2(g)$.
Assume the system comprises $C(s)$, $C_2H_2(g)$, $CH_4(g)$, $CO(g)$, $CO_2(g)$, $H_2(g)$, and $H_2O(g)$.

    (a) Using the Lagrangian multiplier method, evaluate the equilibrium composition, the oxygen potential, and the carbon potential as a function of temperature and $C_2H_2(g)/O_2(g)$ ratio.

    (b) Evaluate the adiabatic flame temperature as a function of the $C_2H_2(g)/O_2(g)$ ratio.

    (c) What molar ratio of $C_2H_2(g)/O_2(g)$ is required in order to avoid C deposition at all temperatures?

18.2 A system is assumed to comprise the species $C(s)$, $CO_2(g)$, $CO(g)$, and $O_2(g)$ at 1000 K.

    (i) Construct the formula matrix.

    (ii) Use the Lagrangian multiplier method to derive a set of independent chemical reactions for this system.

    (iii) Write down the equations for the empirical equilibrium constants assuming perfect gas mixtures.

    (iv) Derive the relation between G and the extents of reaction if the system initially contains 1 mol $C(s)$ and 1 mol $CO_2(g)$ at 2 bars.

    (v) Obtain expressions for the derivatives of G with respect to the extents of reaction.

18.3 Solve this same problem using the Lagrangian multiplier method.

18.4 Derive the expressions for carbon, sulfur, and nitrogen potentials in terms of molecular gas species using the reaction equations discussed in this chapter.

19 Chemical Equilibria Between Gaseous and Condensed Phases I

Chemical reactors in which a complex gas mixture is in contact with a condensed phase are often used in materials production and processing. The thermodynamic problem of interest here is in deciding whether or not the condensed-phase material is carburized, hydrided, oxidized, sulfurized, nitrided, silicided, phosphorized, and so on, by the surrounding complex gas mixture.

We have seen earlier how the thermodynamic data for many pure substances in their standard states are presented in different forms in computer databases as compared with their presentation in tables:

(i) In computer databases the SER state is usually preferred. This involves the storing of several coefficients from which all the thermodynamic properties can be derived.

(ii) In many tabular presentations the SSR state is used. All the necessary differentiations/integrations have already been carried out and the various thermodynamic properties can be read at regular temperature intervals.

19.1 GRAPHICAL PRESENTATION OF STANDARD THERMOCHEMICAL DATA

Computer calculations of complex heterogeneous chemical equilibria can lead to voluminous data outputs so that there is much to be said in favor of a simpler graphical presentation of the results, just as phase diagrams are so useful in the case of phase equilibria. While there is a loss of accuracy in presenting the data this way and hence in any calculations performed in using them, graphical solutions can be very useful for carrying out back-of-the-envelope calculations and obtaining a bird's eye view of the behavior for a class of systems. It is on this aspect of the graphical representation of heterogeneous chemical equilibria which this and the following chapters are concerned.

The graphical representation is made particularly useful from the realization that, for almost all compounds, $\Delta_f G^\circ$ is, to a very good approximation, usually

Materials Thermodynamics. By Y. Austin Chang and W. Alan Oates
Copyright 2010 John Wiley & Sons, Inc.

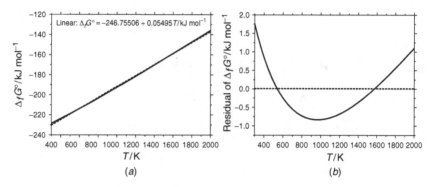

Figure 19.1 Near linearity of $\Delta_f G^\circ(H_2O(g))$ as a function of temperature: (a) JANAF table values for $\Delta_f G^\circ(H_2O(g))$ together with linear representation; (b) residuals between JANAF table values and linear plot values.

a linear function of temperature (in a transformation-free region). This is particularly true when the accuracy of much high-temperature data is taken into account. As an illustration, consider $\Delta_f G^\circ(H_2O(g))$ since the thermodynamic properties for this substance are among the most accurately known. Figure 19.1a shows $\Delta_f G^\circ(H_2O(g))$ as a function of temperature according to data in the JANAF tables. There is a slight nonlinearity in this curve. Also shown (dashed) in the figure is a linear curve obtained from a least-squares fit to the JANAF data. The small deviation from linearity is made more apparent in Figure 19.1b where the residuals of the linear representation from the tabulated values are shown. It can be seen that the differences in $\Delta_f G^\circ(H_2O(g))$ calculated from the linear approximation are only slightly different from the JANAF values. For other substances, where the thermodynamic properties are not known with the same degree of accuracy as for $H_2O(g)$, it is clear that a linear representation is often of sufficient accuracy in most calculations.

The near linearity of $\Delta_f G^\circ - T$ plots is emphasized in Figure 19.2, which shows such a plot for several oxides. In some cases a phase transformation occurs in either metal or oxide causing a change of slope, but, otherwise, the variation with temperature is seen to be essentially linear. We will continue to use linear representations of the $\Delta_f G^\circ - T$ variation for compound formation throughout the remainder of this Chapter.

A table of some linear approximations for $\Delta_f G^\circ - T$ for some compounds which are required in the exercises in this and other Chapters is given in Appendix A.

19.2 ELLINGHAM DIAGRAMS

It was pointed out by Ellingham that there are advantages to be gained by plotting not just $\Delta_f G^\circ - T$, where formation refers to 1 mol of compound, but also

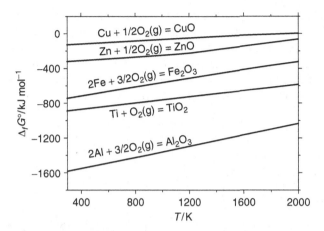

Figure 19.2   The $\Delta_f G^\circ$ as function of temperature for several oxides.

$\Delta_r G^\circ - T$, where the reaction refers to 1 mol of $O_2(g)$. Figure 19.3 shows the same data as plotted previously in Figure 19.2 but now on a per mole of $O_2(g)$ basis rather than per mole of compound.

As we discuss below, the reason for plotting per mole of $O_2(g)$ lies in the ability of being able to use plots in this form for carrying out equilibrium calculations rather than as just a means of presenting thermodynamic data.

Most of the lines shown in Figure 19.4 are for the formation of an oxide from the elements but some are not. What they have in common is that all the reactions refer to 1 mol of $O_2(g)$, for example,

$$2Fe(s) + O_2(g) = 2FeO(s)$$

$$6FeO(s) + O_2(g) = 2Fe_3O_4(s)$$

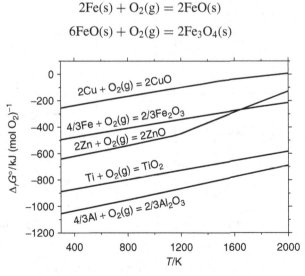

Figure 19.3   Same thermochemical data as used in Figure 19.2 but plotted per mole $O_2(g)$.

A more complete set of $\Delta_r G^\circ - T$ curves for the formation of oxides is shown in Figure 19.4. Similar collections of curves for different groups of compounds—sulfides, carbides, nitrides, chlorides, and so on—have been published. We will continue to concentrate on metal/metal oxide Ellingham diagrams, but the following remarks pertaining to the $\Delta_r G^\circ - T$ curves in Figure 19.4 are equally valid to diagrams for other kinds of compounds:

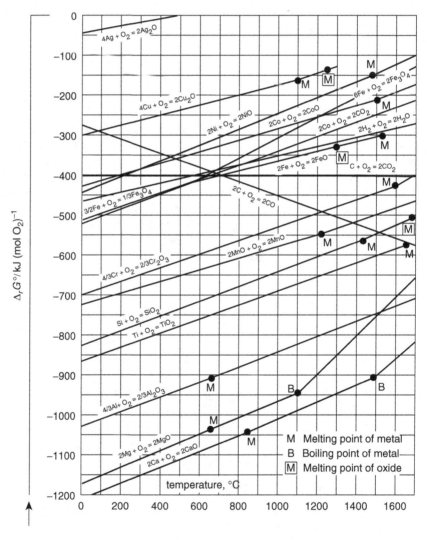

Figure 19.4   Meta/metal oxide Ellingham diagram. Adapted from D.R. Gaskell, *Thermodynamics of Materials*, 3rd Edition, Taylor and Francis, Washington, DC, 2002.

1. The slope of the *tangent* to a curve at any chosen temperature gives the negative of the standard entropy of reaction at that temperature:

$$\Delta_r S^\circ(T) = -\frac{d\Delta_r G^\circ(T)}{dT}$$

2. The intercept at $T = 0$ K of the *tangent* to the curve at T gives $\Delta_r H^\circ(T)$.

3. The curves of $\Delta_r G^\circ(T)$ versus T are almost straight lines between transition points so that a good approximation for $\Delta_r G^\circ(T)$, over a specified temperature range, is

$$\Delta_r G^\circ(T) \approx \langle \Delta_r H^\circ \rangle - T \langle \Delta_r S^\circ \rangle$$

4. Although good estimates of $\Delta_r G^\circ(T)$ can be obtained from the straight-line approximation, it is less accurate to try and obtain $\Delta_r H^\circ(T)$ and $\Delta_r S^\circ(T)$ from $\langle \Delta_r H^\circ \rangle$ and $\langle \Delta_r S^\circ \rangle$; that is, there is some canceling of errors in the sum, $\Delta_r G^\circ$.

5. A majority of the curves in Figure 19.4 are approximately parallel, the occasional one is approximately horizontal and, rarely, a curve may have a negative slope. These patterns in the slopes of the curves can be understood from the equation

$$\Delta_r S^\circ(T) = \sum_i v_i S_i^\circ(T) \qquad (19.1)$$

and the fact that $S^\circ(\text{gases}) \gg S^\circ(\text{solids or liquids})$.

If we now look at the different types of reaction equations whose values are plotted in Figure 19.4, we can then see, using (19.1), that

$$M(s) + O_2(g) = MO_2(s) \qquad\qquad \Delta_r S^\circ \approx -S^\circ(g)$$
$$2H_2(g) + O_2(g) = 2H_2O(g) \qquad\qquad \Delta_r S^\circ \approx -S^\circ(g)$$
$$C(s) + O_2(g) = CO_2(g) \qquad\qquad \Delta_r S^\circ \approx 0$$
$$2C(s) + O_2(g) = 2CO_2(g) \qquad\qquad \Delta_r S^\circ \approx +S^\circ(g)$$

6. Changes in the slope are found when phase transformations occur in either the reactant or product phases. These can be understood by considering the melting of either the reactant or product:

$$M(s) + O_2(g) = MO_2(s) \qquad \Delta_r G^\circ(T) = -A + BT$$
$$M(s) = M(l) \qquad \Delta_{fus} G^\circ(T) = C - DT$$
$$MO_2(s) = MO_2(l) \qquad \Delta_{fus} G^\circ(T) = E - FT$$
$$M(l) + O_2(g) = MO_2(s) \qquad \Delta_r G^\circ(T) = -(A + C) + (B + D)T$$
$$M(s) + O_2 = MO_2(l) \qquad \Delta_r G^\circ(T) = -(A - E) + (B - F)T$$

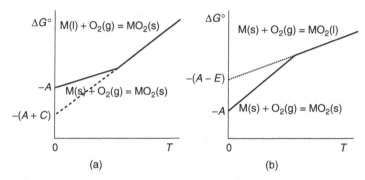

Figure 19.5   Variations in slope caused by phase transformations in reactant and product phases: (*a*) increase in slope occurring at phase transformation occurring in reactant phase; (*b*) decrease in slope occurring at phase transformation occurring in product phase.

As is apparent from these equations and from Figure 19.5*a*, there is an *increase* in slope with T when a reactant condensed phase undergoes a phase transition and, in Figure 19.5*b*, a *decrease* in slope with T when a product condensed phase undergoes a phase transition.

Since $\Delta_{vap}S° \gg \Delta_{fus}S° > \Delta_{\alpha}^{\beta}S°$, these changes in slope are much greater at vaporization transitions than at fusion.

Although the above discussion relates to a discussion of metal/metal oxide systems, the same principles apply to the other types of Ellingham diagrams.

19.2.1   Chemical Potentials

We now discuss why Ellingham diagrams are of value when examining heterogeneous chemical equilibria. In the next chapter we will emphasize that this statement comes with an extremely important proviso *when they are used correctly*. It is very easy to use them incorrectly.

Concentrating on the metal/metal oxide Ellingham diagram, the equivalence of $\Delta_r G°(T)$ with the oxygen chemical potential is easily demonstrated by considering the reaction

$$M(s) + O_2(g) = MO_2(s) \tag{19.2}$$

for which we may write

$$\Delta_r G = \Delta_r G° + RT \log_e Q_p$$

where Q_p is the reaction equation quotient previously defined in (17.4). Since the metal and metal oxide being considered are pure, this last equation can be rewritten as

$$\Delta_r G = \Delta_r G° - RT \log_e p_{O_2} \quad \text{perfect gas, pure condensed} \tag{19.3}$$

At equilibrium $\Delta_r G = 0$ so that, for any p_{O_2},

$$\boxed{\Delta_r G^\circ = RT \log_e p_{O_2}^{eq} = \Delta\mu_{O_2}^{eq}} \quad \text{perfect gas, pure condensed} \quad (19.4)$$

The term oxygen chemical potential is usually abbreviated to oxygen potential and we will use this terminology here. Similarly, we can speak of carbon potential.

Equation (19.4) means that the ordinate in Figure 19.4, $\Delta_r G^\circ$, can equally well be regarded as the oxygen potential, $\Delta\mu_{O_2}^{eq}$, when equilibrium between gas and pure condensed phases is being discussed. Away from equilibrium, we can use (19.3) and (19.4) to obtain

$$\Delta_r G = RT \log_e p_{O_2}^{eq} - RT \log_e p_{O_2} \quad (19.5)$$

which may be rewritten as

$$\boxed{\Delta_r G = \Delta\mu_{O_2}^{eq} - \Delta\mu_{O_2}} \quad (19.6)$$

From this last equation we can see that:

(i) If $\Delta\mu_{O_2} > \Delta\mu_{O_2}^{eq}$ or, equivalently, $p_{O_2} > p_{O_2}^{eq}$, then $\Delta_r G < 0$, which means that the reaction equation given in (19.2) favors *product* formation.

(ii) If $\Delta\mu_{O_2} < \Delta\mu_{O_2}^{eq}$ or, equivalently, $p_{O_2} < p_{O_2}^{eq}$, then $\Delta_r G > 0$, which means that the reaction equation given in (19.2) favors *reactant* formation.

This information can be included on a $\Delta\mu_{O_2} - T$ diagram, as shown in the sketch given in Figure 19.6. Equilibrium between all three phases (two solids, one gas) only exists along the heavy $\Delta_r G^\circ$ line. The dashed line intersects the heavy line at a temperature where $\Delta_r G = 0$. At a lower temperature than this, $\Delta_r G < 0$

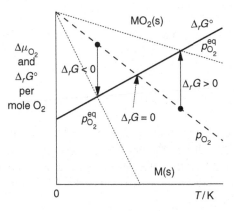

Figure 19.6 Driving Forces on a $\Delta\mu_{O_2}$ versus T diagram.

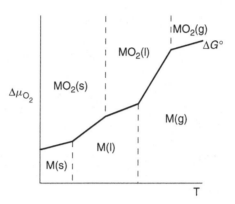

Figure 19.7   Domains of stability for various phases.

(product phase stable), the equilibrium p_{O_2} at this temperature being given by the intersection of the dotted line with the heavy line. Similar reasoning follows for temperatures higher than those corresponding with the original intersection, where the reactant phase is now stable.

Summarizing: at oxygen potentials higher than those corresponding to the heavy line, the product phase is stable, the reactant phase unstable. The reverse is true for lower oxygen potentials than those corresponding to the heavy line.

This recognition of the importance of driving force leads to the concept of domains of stability for the different phases. These can be indicated on an Ellingham diagram. One is shown schematically in Figure 19.7. Note that, if M(g) or MO_2(g) are involved, the lines on the diagram refer to these phases at the standard pressure so that the marked domains of stability also refer to these conditions.

EXERCISES

19.1   One step in the manufacture of specially purified nitrogen is the removal of small amounts of residual oxygen by passing the gas over copper gauze at approximately 500°C. The following reaction takes place:

$$Cu(s) + \tfrac{1}{2}O_2(g) = Cu_2O(s)$$

(a) Assuming that equilibrium is reached in this process, calculate the oxygen pressure in the purified nitrogen.

(b) What would be the effect of raising the temperature to 800°C? What if the temperature is lowered to 300°C? What is the probable reason for using 500°C?

(c) What would be the effect of increasing the pressure?

19.2 In a gas analysis train which is part of a vacuum fusion apparatus, carbon monoxide is oxidized to carbon dioxide by cupric oxide (which is subsequently reduced to cuprous oxide) at about $300°C$.

(a) Calculate the concentration of residual carbon monoxide in equilibrium with carbon dioxide, assuming the latter to be at 0.5 bar pressure.

(b) In the operation of this equipment, the gases are passed through an efficient adsorber of carbon dioxide and recirculated over the cupric oxide. How does this affect the residual carbon monoxide concentration?

19.3 The partial pressure of oxygen in equilibrium with $Cu_2O(s)$ and CuO has been found to be 0.028 bar at $900°C$ and 0.1303 bar at $1000°C$. Calculate $\Delta_r H°$ for the reaction

$$2Cu_2O(s) + O_2(g) = 4CuO(s)$$

at 298, 1173, and 1273 K.

19.4 Hydrogen is often used to protect Cr from oxidation at high temperatures. However, $H_2(g)$ always contains trace amounts of $H_2O(g)$.

(a) How much moisture content in $H_2(g)$ can be tolerated at 1300 K before chromium becomes oxidized?

(b) Is the oxidation of Cr by H_2O endothermic or exothermic at 1300 K? What, if anything, can be said from this reaction about the heat effect at 1300 K by the oxidation of Cr by pure O_2?

(c) Would the equilibrium in the above reaction be affected by a change in the pressure of the H_2–H_2O mixture from 1 bar to 2 bars? What if it were changed to 200 bars?

	$\Delta_f H°$(298 K)/kJ mol^{-1}	gef/J mol^{-1} K^{-1}
$H_2(g)$	0	130.68
$H_2O(g)$	−241.8	188.83
$Cr(s)$	0	23.64
$Cr_2O_3(s)$	−1134.7	81.17

19.5 A mixture of argon gas and hydrogen gas at 1 bar total pressure is passed through a reaction chamber containing a mixture of liquid Sn and liquid $SnCl_2$ at 900 K. The composition of the gas leaving the reaction chamber is 50% H_2, 0.01% HCl, and 49.99% Ar. Has equilibrium been attained between the gas phase and the liquid phase in the reaction chamber?

19.6 A solid metal M can form two oxides, MO and M_3O_4. The metal can exist in equilibrium with one of these oxides at low temperatures and can exist in equilibrium with the other oxide at high temperatures. The Gibbs energies of formation of these oxides are

$$\Delta_f G^\circ (MO(s), T)/kJ \ mol^{-1} = -259.6 + 0.06255T$$

$$\Delta_f G^\circ (M_3O_4(s), T)/kJ \ mol^{-1} = -1090.8 + 0.3128T$$

Determine which of the two oxides is in equilibrium with the metal at room temperature and the maximum temperature at which this oxide is in equilibrium with M.

19.7 Plot %CO_2 as a function of temperature from 400 to 1200°C for the oxidation of:

(a) Fe to magnetite (Fe_3O_4) at low temperatures and to wustite (FeO) at high temperatures

(b) wustite (FeO) to magnetite (Fe_3O_4)

(c) magnetite (Fe_3O_4) to hematite (Fe_2O_3)

Would you use CO_2 to oxidize magnetite to hematite (Fe_2O_3) at these temperatures?

19.8 Calculate the equilibrium oxygen pressure between Al_2O_3 and Al(l) at 1000 K. Could a vacuum of 10^{-10} torr prevent the oxidation of Al?

20 Chemical Equilibria Between Gaseous and Condensed Phases II

In the previous chapter, we introduced the graphical presentation of $\Delta_f G^\circ$ and $\Delta_r G^\circ$ for a selected group of compounds, for example, oxides and sulfides. It was noted that, when phase transitions do not interfere, these two thermodynamic quantities are, usually, nearly linear functions of temperature. In practice, a linear representation of $\Delta_f G^\circ$ versus T is often within the accuracy of the experimental results. Some examples of such representations are given in Appendix A, and these can be used in many of the exercises in this chapter and Chapter 21.

We also noted that a graphical presentation of standard Gibbs energies which is of particular value is the one introduced by Ellingham where, instead of plotting $\Delta_f G^\circ$ versus T, a graph of $\Delta_r G^\circ$ versus T is plotted, where all the members of a group of reaction equations refer to 1 mol of $O_2(g)$ (or, e.g., S_2). The advantage of this method of plotting comes from the fact that, when the condensed phases are in their pure state, then, from the relation between the empirical equilibrium constant (for the metal/metal oxide case),

$$\Delta_r G^\circ \text{ per mole } O_2(g) = RT \log_e p_{O_2} = \Delta\mu_{O_2} \tag{20.1}$$

In this case, all the domains of stability for the condensed-phase substances presented on the Ellingham diagram are categorized in terms of the oxygen potential. By plotting all reaction equations on the basis of 1 mol of $O_2(g)$, the graph has been transformed from one of just being the graphical presentation of thermochemical data to one of value in carrying out equilibrium calculations.

Ellingham diagrams are particularly useful for back-of-the-envelope calculations and for visualizing heterogeneous chemical equilibria. They can, however, be used badly and we will pay particular attention to this aspect.

For illustration purposes, we will mainly concentrate on metal/metal oxide diagrams, but the same principles apply to Ellingham diagrams for all the other groups of compounds.

Materials Thermodynamics. By Y. Austin Chang and W. Alan Oates
Copyright 2010 John Wiley & Sons, Inc.

20.1   SUBSIDIARY SCALES ON ELLINGHAM DIAGRAMS

The value of Ellingham diagrams can be enhanced by the inclusion of various subsidiary scales placed around the perimeter of a diagram. We can illustrate this by first considering how $\Delta\mu_{O_2}$ varies for a fixed value of p_{O_2}.

Using (20.1), this variation is illustrated in Figure 20.1 for $p_{O_2} = 10^{-10}$ bar. At 2000 K, the point on the figure indicates a value of $\Delta\mu_{O_2} = -382.9$ kJ (mol O_2)$^{-1}$ for this oxygen pressure.

It can be seen that the whole diagram can be covered by an oxygen pressure fan, with origin at zero oxygen potential and 0 K. In order to be able to have a fairly accurate idea of p_{O_2} at any point on the diagram, it is convenient to add a nomogram around the perimeter of the diagram. This enables the connection between p_{O_2}, $\Delta\mu_{O_2}$, and T to be read immediately at any point on the diagram. The construction of the nomogram is illustrated in Figure 20.1b. Note that the anchor point at 0 K is at $\Delta\mu_{O_2} = 0$ for all values of p_{O_2}.

It is also possible to include information about the oxygen potentials of complex gas mixtures on the $\Delta\mu_{O_2}$–T diagram. In Chapter 18 we saw how knowledge of $\Delta_r G^\circ$ for the reaction equations

$$2CO(g) + O_2(g) = 2CO_2(g)$$

$$2H_2(g) + O_2(g) = 2H_2O(g)$$

could be used to define $\Delta\mu_{O_2}$ in terms of the (p_{CO_2}/p_{CO}) or (p_{H_2O}/p_{H_2}) ratios:

$$\Delta\mu_{O_2} = 2\left[\Delta_f G^\circ(CO_2(g), T) - \Delta_f G^\circ(CO(g), T)\right]$$

$$+ 2RT \log_e\left(\frac{p_{CO_2}}{p_{CO}}\right) \qquad (20.2)$$

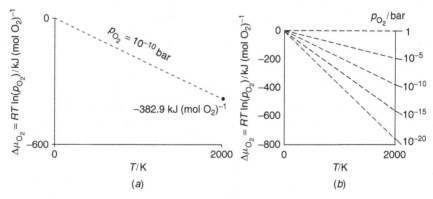

Figure 20.1 Construction of p_{O_2} nomogram on metal/metal oxide Ellingham diagram: (a) oxygen potential as function of T for $p_{O_2} = 10^{-10}$; (b) fan of lines for various p_{O_2}.

$$\Delta\mu_{O_2} = 2\,\Delta_f G^{\circ}(H_2O(g), T) + 2RT\,\log_e\left(\frac{p_{H_2O}}{p_{H_2}}\right) \qquad (20.3)$$

It can be seen that there are fans of lines for the oxygen potentials obtainable from gas mixtures containing different ratios of these molecular species. Showing the dependence of oxygen potential on these ratios is again most conveniently done with the aid of nomograms, comprising an anchor point for the fan at 0 K together with a perimeter scale. Figure 20.2 shows how the oxygen potential can be specified at any T through p_{O_2}, p_{CO_2}/p_{CO}, or p_{H_2O}/p_{H_2}.

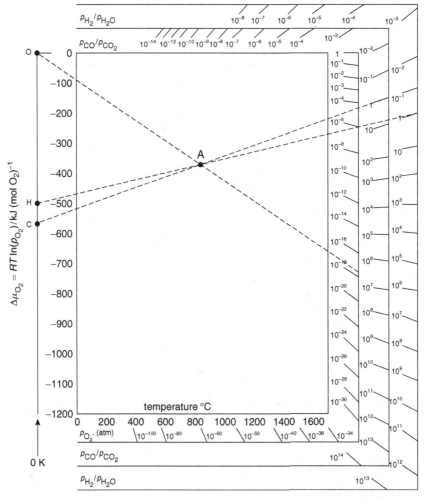

Figure 20.2 $\Delta\mu_{O_2}$ versus T, showing how oxygen potential can be specified in terms of gas mixture composition. Adapted from D. R. Gaskell, *Thermodynamics of Materials*, Taylor and Francis, Washington, DC, 2002.

With the aid of the subsidiary scales, it is now possible to assign values to T, p_{O_2}, $\Delta\mu_{O_2}$, as well as p_{CO_2}/p_{CO} or p_{H_2O}/p_{H_2} ratios. The dashed lines in Figure 20.2 are for $p_{H_2O}/p_{H_2} = p_{CO_2}/p_{CO} = 1$. They correspond with $\Delta_r G^\circ$ for both reactions. The point marked A in the figure is where a gas mixture containing equal amounts of all four gases can exist and is known as the water gas equilibrium. The temperature is $\approx$1095 K and the oxygen pressure there is $\approx$5 × 10^{-18} bar, a quite low oxygen potential but one easily controlled using gas mixtures containing the four molecular species under consideration.

Figure 20.3 shows how the database for $\Delta_r G^\circ$ /kJ mol O_2^{-1} shown in Figure 19.4 can be combined with the plot given in Figure 20.2 for $\Delta\mu_{O_2}$/kJ mol O_2^{-1},

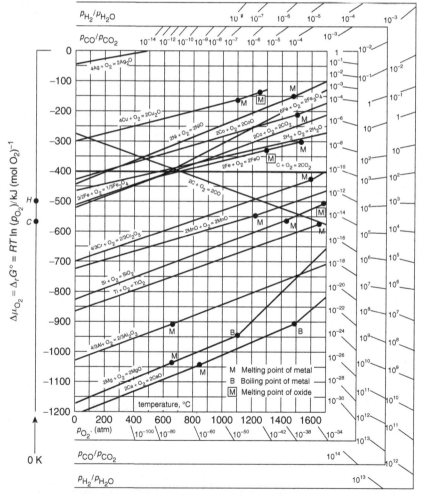

Figure 20.3 Metal/metal Oxide Ellingham diagram. Adapted from D. R. Gaskell, *Thermodynamics of Materials*, 3rd Edition, Taylor and Francis, Washington, DC, 2002.

to yield a very useful way of carrying out calculations of heterogenous chemical equilibria. Although we have concentrated our discussion on oxide equilibria, analogous diagrams can be constructed for sulfides, nitrides, carbides, halides, etc. Provided that these diagrams are used correctly (we illustrate later how it is very easy to reach wrong conclusions when Ellingham Diagrams are used incorrectly), they provide a very useful means for quickly obtaining approximate answers for equilibrium calculations.

It is worth noting that, if the oxygen potential is being controlled by using CO_2/CO mixtures (or by any other gas mixture), it is unnecessary for $O_2(g)$ to be actually present in the gas phase to obtain the desired oxygen potential. It may or may not be actually present (usually in immeasurably small concentrations when it is), this being a function of the kinetics of the reactions involved. From a thermodynamic point of view, this is of no consequence; only the CO_2/CO ratio at the selected temperature is important.

This example of using a gas mixture to control an elemental chemical potential when the actual element itself is not present (a metastable equilibrium) can be particularly well illustrated by another example. A mixture of $H_2(g)$ and $NH_3(g)$ can be used to control the nitrogen potential of the gas phase. The appropriate reaction equation is

$$3H_{2(g)} + N_{2(g)} = 2NH_{3(g)}$$

from which we can see that

$$\Delta\mu_{N_2} = \Delta_r G^\circ + RT \log_e \left(\frac{(p_{NH_3})^2}{(p_{H_2})^3} \right) \tag{20.4}$$

At high temperatures, however, $\Delta_r G^\circ$ for this reaction is positive and, as a result, $NH_3(g)$ should start to decompose to $N_2(g)$ and $H_2(g)$. This would mean that the gas ratio and hence the nitrogen potential would be quite different from that intended. But the decomposition of $NH_3(g)$ requires a catalyst and may not occur. If the decomposition can be avoided, a gas mixture of $H_2(g)$ and $NH_3(g)$ can actually be used to obtain high nitrogen potentials; this is taken advantage of in the nitriding of steel, summarized in the reaction equation

$$3H_2(g) + 2N(\text{in Fe}) = 2NH_3(g)$$

20.2   SYSTEM DEFINITION

Having said that a primary concern of thermodynamics is in predicting properties of systems in equilibrium, we emphasize again that *the application of thermodynamics will provide the wrong output if it is fed the wrong input. It is absolutely essential that the system should be defined appropriately.*

We introduced the important point in Chapter 18 that the responsibility for defining the system lies in the hands of the thermodynamicist. A simple adiabatic flame temperature calculation in the combustion of hydrogen with oxygen was used as an example. In systems comprising only a gas phase, poor system definition is rare these days since all relevant species will normally be contained in the computer database used. This may not be the case, however, when the system involves condensed phases since the thermodynamicist may, sometimes, be unaware of all the phases which should be taken into account. In such cases, it is very easy to carry out calculations for a metastable equilibrium in a poorly defined system instead of on the true equilibrium for a properlydefined system.

Relying on the information presented in Ellingham diagrams can be particularly dangerous in this regard because many of the lines presented refer to metastable equilibria and not all known species occurring in a particular metal/metal oxide (e.g., sulfide) system are taken into account. We will illustrate these important points by considering two examples.

Example 20.1 The Fe–O System

Figure 20.4 shows $\Delta_r G^\circ$ per mole O_2 as a function of temperature for the six possible reaction equations which involve Fe(s), FeO(s), Fe_3O_4(s), and Fe_2O_3(s); that is, these lines are a mini–Ellingham diagram. The kinks in the curves are where phase transformations occur. The temperature–composition phase diagram for this system is shown in Figure 20.5. From this we can immediately see which of the six curves in Figure 20.4 represent metastable equilibria and which are for the stable equilibria. It follows that the oxygen potential–temperature relation for this system is not that presented in our mini–Ellingham diagram, Figure 20.4. We require a phase diagram in oxygen potential–temperature space analogous to the $T-x_O$ phase diagram given in Figure 20.5. In order to distinguish the Ellingham diagram from the equilibrium phase diagram, we plot T versus μ_{O_2} rather than

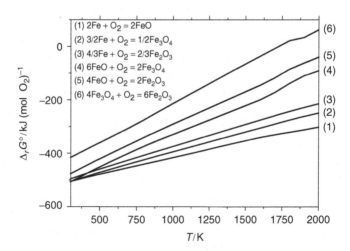

Figure 20.4 Standard Gibbs energies of reaction per mole of O_2 for various reactions in Fe–O system.

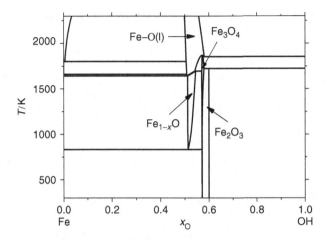

Figure 20.5 $T-x_O$ phase diagram for Fe–O system.

the reverse. This mode of plotting, shown in Figure 20.6, also has the advantage that its appearance is more like the more familiar $T-x_O$ phase diagram. It is also the method of presentation introduced earlier in Chapter 11.

The phase diagram given in Figure 20.6 may safely be used in true equilibrium calculations in the Fe–O system. It could, of course, incorporate subsidiary scales for p_{O_2}, p_{CO_2}/p_{CO}, or p_{H_2O}/p_{H_2} just as on the Ellingham diagram.

Example 20.2 Melting of Ti in Al$_2$O$_3$ Crucibles

A good example illustrating poor system definition by the thermodynamicist concerns the feasibility of melting Ti metal in an alumina (Al$_2$O$_3$) crucible with

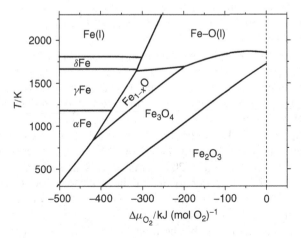

Figure 20.6 $T-\mu_{O_2}$ phase diagram for Fe–O system.

the aim of producing ductile Ti, a property which depends on the metal having a very low O content.

We will proceed as we did in calculating the adiabatic flame temperature in Chapter 18; namely, we make three different assumptions, of increasing sophistication concerning the definition of the system. The standard melting point of Ti is 1943 K. We will concentrate on calculations at 2200 K (chosen so that Ti–O alloys are liquid at all concentrations of O):

(i) *Thermodynamicist A*   defines the system as comprising Ti(l), Al_2O_3(s), Al(l), and TiO_2(s), all pure. Consequently, the following chemical reaction equation is relevant:

$$\tfrac{3}{2}Ti(l) + Al_2O_3(s) = 2Al(l) + \tfrac{3}{2}TiO_2(s) \tag{20.5}$$

Thermodynamicist A uses the Ellingham diagram in Figure 20.3 for his or her source of data.

Consider any pair of reactions on that diagram that evaluate the Gibbs reaction energy for a condensed-phase reaction equation such as

$$A(s) + BO_2(s) = B(s) + AO_2(s)$$

Then $\Delta_r G^\circ\ (T)$ for this reaction is given by

$$\Delta_r G^\circ(T) = \Delta_f G^\circ(AO_2(s), T) - \Delta_f G^\circ(BO_2(s), T)$$
$$= \Delta\mu_{O_2}^{eq}(A/AO_2) - \Delta\mu_{O_2}^{eq}(B/BO_2)$$

and we see that $\Delta_r G^\circ < 0$ when $\Delta_f G^\circ(AO_2(s), T) < \Delta_f G^\circ(BO_2(s), T)$. Put another way: If the A/AO_2 line is lower down on the Ellingham diagram than the B/BO_2 line, then A(s) will reduce BO_2(s) and vice versa.

Using Figure 20.3, thermodynamicist A concludes that, since the Al/Al_2O_3 line lies below that for Ti/TiO_2 on the Ellingham diagram, it is safe to melt Ti in an Al_2O_3 crucible.

(ii) *Thermodynamicist B*   is aware that, in order for Ti to be technologically useful, it must possess a very low oxygen concentration and is also aware, as is apparent in Figure 20.7, that both solid and liquid Ti can absorb considerable amounts of oxygen. It is also clear from this phase diagram that the coexistence of Ti with TiO_2, as might be implied from using Figure 20.3, is very much a metastable equilibrium. The important thing to ask is what oxygen potential is required to melt Ti with a low oxygen content? He or she calculates the T–μ_{O_2} phase diagram shown in Figure 20.8. This reveals that extremely low oxygen potentials are required in order to obtain the necessary low O concentrations in liquid Ti at 2200 K. These low oxygen potentials are compared with the oxygen potentials associated with Al/Al_2O_3 in Figure 20.9. According to the Ellingham diagram calculation by thermodynamicist A, the fact that the Ti/TiO_2 curve lies above (at higher oxygen potentials) than the Al/Al_2O_3 curve, leads

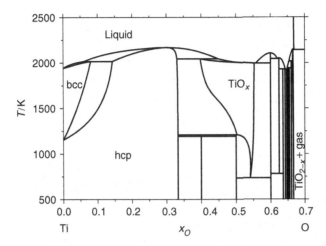

Figure 20.7 Temperature–composition diagram for Ti–O system. The black area comprises several closely spaced intermediate phases.

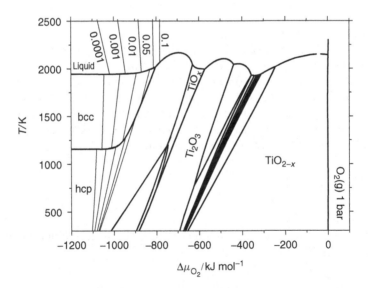

Figure 20.8 Temperature–oxygen potential diagram for Ti–O system.

thermodynamicist A to conclude that it is safe to melt Ti in an Al_2O_3 crucible. It can be seen from Figure 20.9 that the oxygen potential required to obtain O-free Ti is approximately -1100 kJ $(mol\ O_2)^{-1}$, which is much lower than that required for the stability of $Al_2\ O_3$, as can be seen from the location of the Al/Al_2O_3 curve. Thermodynamicist B concludes, correctly, that it is not possible to obtain O-free Ti by melting in Al_2O_3 crucibles.

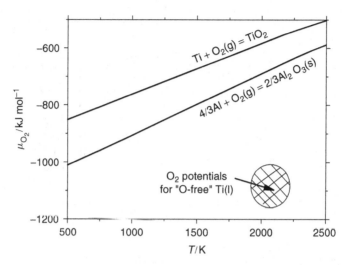

Figure 20.9 Oxygen potentials required for obtaining oxygen-free Ti(l) are shown circled.

(iii) *Thermodynamicist C*   realizes that thermodynamicist B's system definition is much better that thermodynamicist A's, who reached an incorrect conclusion resulting from a very poor system definition. Although thermodynamicist B reached the correct conclusion, thermodynamicist C realizes that it has been reached via a system definition which is still not complete: It is necessary to consider all possible phases occurring in the ternary Ti–Al–O system.

Since we are only considering binary systems in this chapter, discussion of thermodynamicist C's deliberations is postponed until Chapter 22.

EXERCISES

20.1 The following reactions are sufficient to evaluate the equilibrium in a C(s), $O_2(g)$, $CO(g)$, $CO_2(g)$ system:

$$CO(g) + \tfrac{1}{2}O_2(g) = CO_2(g) \tag{20.6}$$

$$C(s) + O_2(g) = CO_2(g) \tag{20.7}$$

Obtain $\Delta_r G^\circ$ for reactions (1) and (2) from the oxide Ellingham diagram shown in Figure 20.3.

(a) If we start with 1 mol CO(g) and feed it into a closed quartz box maintained at a total external pressure of 1 bar and 1000 K, evaluate the final partial pressures of the gases.

(b) Repeat the calculation for the case where the box is made of Fe and also calculate how much C is formed. *Note*: You will simplify the calculation in this part considerably if you use the metal oxide Ellingham diagram to help you make a good approximation.

Important Note: Carbon deposition from CO(g) requires a catalyst before it can occur. Iron is one such catalyst, quartz is not.

20.2 (a) Draw a carbon potential $\Delta\mu_C(T)$–T diagram. Make the T axis 0–2000 K and the ordinate axis run from $+200$ to -200 kJ (mol C)$^{-1}$.

(b) Put on a fan (anchor point and perimeter scale) which enables specification of $\Delta\mu_C(T)$ in terms of a combination of CH_4 and H_2 partial pressures.

(c) Put on a fan (anchor point and perimeter scale) which enables specification of $\Delta\mu_C(T)$ in terms of a combination of CO and CO_2 partial pressures.

(d) Assume a gas containing equimolar amounts of CO and CO_2 at 1 bar total pressure is passed into a furnace at 1300 K:

   (i) Will the C(s) either deposit (assuming the furnace structure contains lots of Fe) in the furnace or exit from the furnace?

   (ii) Will pure Si(s) in the furnace be converted and, if so, to what?

20.3 Calculate the equilibrium oxygen pressure between Al_2O_3 and Al(l) at 1000 K. Would a vacuum of 10^{-8} Pa prevent the oxidation of Al at this temperature?

21 Thermodynamics of Ternary Systems

In the earlier chapters the focus has been on the thermodynamics of unary and binary systems. However, multicomponent systems are often encountered in MS&E. Engineering alloys, for example, usually consist of more than 2 elements, sometimes more than 10. In this and the following chapter, we extend our discussion of the thermodynamics of binaries to that of ternaries. Once the thermodynamic properties of ternaries are understood, it is usually straightforward to extend the concepts to higher order alloys. Experience shows that once all the binary and ternary thermodynamic properties are known, extrapolation to higher order systems can usually be carried out with confidence. This is because the number of near neighbor contacts which involve more than three different types of atoms will usually be very small in multicomponent alloys. Rarely, when strong chemical interactions are present, this can break down due to the formation of multicomponent compound phases.

In line with our earlier discussions of binary alloys, we will first discuss the thermodynamics of ternary solution phases and then discuss phase equilibria. More general aspects of the geometric presentation of phase equilibria in ternary alloys, including heterogeneous chemical equilibria, are discussed in the next chapter.

We have seen that, for two *preselected* phases α and β, the conditions of equilibrium, assuming all species are mobile between both phases, are given by

$$\mu_i^\alpha = \mu_i^\beta \quad \forall\, i \tag{21.1}$$

In order to be able to apply (21.1) to ternary phase equilibria we need to know the μ_i in the participating ternary phases. Because of the relatively poor availability of experimental results for ternary systems, the estimation of the thermodynamic properties of ternaries from those of the constituent binaries takes on a cardinal role and we first discuss some ways by which this is done.

Materials Thermodynamics. By Y. Austin Chang and W. Alan Oates
Copyright 2010 John Wiley & Sons, Inc.

21.1 ANALYTICAL REPRESENTATION OF THERMODYNAMIC PROPERTIES

21.1.1 Substitutional Solution Phases

It is necessary to decide which composition of a binary alloy should be used in estimating the contribution of the properties from that binary to the properties of a selected ternary alloy. Figure 21.1 shows just four of an infinite number of possible ways which could be used in deciding which composition of the binary BC would be the most appropriate for estimating the contribution of this binary's property to the ternary alloy's property (at the circled point). Two widely used symmetric geometric methods for adding the binary contributions to the ternary, due to Kohler and Muggianu, respectively, are shown in Figure 21.2. An asymmetric addition may be more appropriate in other circumstances, for example, if one of the components is an interstitial element. Here we will concentrate on symmetric extrapolations for substitutional solutions.

If both the binaries and the ternary are assumed to be regular solutions and also that the various binary interaction energies are the same in the ternary as they are in the relevant binary system, then it is straightforward to calculate the ternary alloy's properties. In this case, the numbers of A–B, B–C, and C–A contacts in the A–B–C ternary are known because of the random mixing assumption. The ternary properties are given by

$$G_m^E(x_A, x_B, x_C) = x_A x_B L_0^{AB} + x_B x_C L_0^{BC} + x_C x_A L_0^{CA} \qquad (21.2)$$

This estimation is just that expected from the Muggianu extrapolation shown in Figure 21.2*b*. This then suggests an obvious way to extend (21.2), namely,

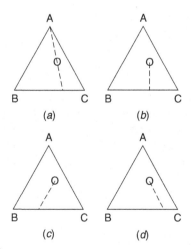

Figure 21.1 Some of the infinite number of possible ways of estimating the B–C contribution to the properties of a ternary A–B–C alloy.

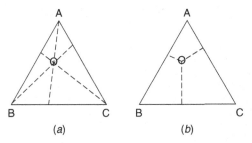

Figure 21.2   Example of two different ways for estimating ternary system properties from constituent binary system properties: (*a*) Kohler extrapolation; (*b*) Muggianu extrapolation.

use the Redlich–Kister polynomial to allow for any deviations from the regular solution model in the binary systems and add them as in (21.2). This assumption leads to the so-called Redlich–Kister–Muggianu (RKM) equation, which is the most widely used extrapolation equation for substitutional and liquid alloys:

$$G_m^E(x_A, x_B, x_C) = x_A x_B \sum_{i=0}^{n} L_i^{AB}(x_A - x_B)^i$$

$$+ x_B x_C \sum_{i=0}^{n} L_i^{BC}(x_B - x_C)^i$$

$$+ x_C x_A \sum_{i=0}^{n} L_i^{CA}(x_C - x_A)^i \tag{21.3}$$

Equations for μ_A^E can be obtained from this equation by using the framed equation given below, which is a general equation for obtaining partial molar quantities from integral quantities for multicomponent systems. Its derivation (we use i for the component of interest and j for any component) is as follows:

$$\mu_i^E = \left(\frac{\partial G^E}{\partial N_i} \right)_{p,T,\{N_j\}} = \left(\frac{\partial N G_m^E}{\partial N_i} \right)_{p,T,\{N_j\}} \tag{21.4}$$

where $N = \sum_i N_i$ and $j \neq i$.

$$\frac{\partial N G_m^E}{\partial N_i} = G_m^E + N \left(\frac{\partial G_m^E}{\partial N_i} \right)$$

$$= G_m^E + N \left[\frac{\partial G_m^E}{\partial x_i} \cdot \frac{\partial x_i}{\partial N_i} + \sum_{j \neq i} \frac{\partial G_m^E}{\partial x_j} \cdot \frac{\partial x_j}{\partial N_i} \right]$$

The relations between the mole fractions and the amounts can be obtained from the differentiation of

$$x_i = 1 - \sum_j \frac{N_j}{N} \tag{21.5}$$

and they are given by

$$\frac{\partial x_i}{\partial N_i} = \frac{1 - x_i}{N}$$

$$\frac{\partial x_i}{\partial N_j} = \frac{-x_i}{N} \qquad j \neq i$$

Finally we can obtain the relation between the partial molar quantity and the integral quantity:

$$\mu_i^E = G_m^E + (1 - x_i)\left(\frac{\partial G_m^E}{\partial x_i}\right) - \sum_{j \neq i} x_j \left(\frac{\partial G_m^E}{\partial x_j}\right)$$

$$\boxed{\mu_i^E = G_m^E + \left(\frac{\partial G_m^E}{\partial x_i}\right) - \sum_j x_j \left(\frac{\partial G_m^E}{\partial x_j}\right)} \tag{21.6}$$

This equation represents a generalization of the equation corresponding with the tangent–intercept method discussed earlier for binary solutions. The equation for binaries can be obtained directly from this general equation, (21.6).

The excess chemical potentials may be obtained from (21.3) for G_m^E by using (21.6). For component A:

$$\mu_A^E = L_0^{AB} x_B (1 - x_A) + L_0^{CA} x_C (1 - x_A)$$

$$+ \sum_{i=1}^n L_i^{AB} x_B (x_A - x_B)^{(i-1)} [(1 - x_A)(x_A - x_B)(i + 1) + i x_B]$$

$$+ \sum_{i=1}^n L_i^{CA} x_C (x_C - x_A)^{(i-1)} [(1 - x_A)(x_C - x_A)(i + 1) + i x_C]$$

$$- x_B x_C \left[L_0^{BC} + \sum_{i=1}^n L_i^{BC} (x_B - x_C)^i (i + 1) \right] \tag{21.7}$$

Similar relations can be written for μ_B^E and μ_C^E.

If extra parameters are required in the fitting to experimental results, the RKM equation can be extended by adding ternary interaction terms to (21.3). These extra terms must be selected so that they reduce to the correct binary solution equation when there are only two components. If we consider just the first two ternary interaction terms, the additional contribution to G_m^E is given by

$$
\begin{aligned}
G_m^E(x_A, x_B, x_C) = x_A x_B x_C L_0^{ABC} &+ \tfrac{1}{3}(1 + 2x_A - x_B - x_C)L_1^{ABC} \\
&+ \tfrac{1}{3}(1 + 2x_B - x_C - x_A)L_1^{BCA} \\
&+ \tfrac{1}{3}(1 + 2x_C - x_A - x_B)L_1^{CAB}
\end{aligned}
\tag{21.8}
$$

and the extra contribution to μ_A^E is given by

$$
\mu_A^E = x_B x_C (1 - 2x_A) L_0^{ABC} + L_1^{ABC}
\tag{21.9}
$$

with analogous equations for the extra contributions to μ_B^E and μ_C^E.

21.1.2   Sublattice Phases

We have already met, in Chapter 14, the following modeling equation for a binary sublattice phase with two equally sized sublattices and whose configurational energy is determined by nearest neighbor interaction energies. In the BW approximation,

$$
\begin{aligned}
\Delta_{\mathrm{mix}} A_m^{\mathrm{conf}} = \frac{z}{2} W_{AB} \left(y_A^\alpha y_B^\beta + y_B^\alpha y_A^\beta \right) \\
+ \sum_\alpha f^\alpha \sum_i y_i^\alpha \log_e y_i^\alpha
\end{aligned}
\tag{21.10}
$$

Assuming there are no additional ternary terms, this can be extended for use with ternary alloys in a manner similar to that used for substitutional phases:

$$
\begin{aligned}
\Delta_{\mathrm{mix}} A_m^{\mathrm{conf}} = \frac{z}{2} W_{AB} \left(y_A^\alpha y_B^\beta + y_B^\alpha y_A^\beta \right) \\
+ \frac{z}{2} W_{BC} \left(y_B^\alpha y_C^\beta + y_C^\alpha y_B^\beta \right) \\
+ \frac{z}{2} W_{CA} \left(y_C^\alpha y_A^\beta + y_A^\alpha y_C^\beta \right) \\
+ \sum_\alpha f^\alpha \sum_i y_i^\alpha \log_e y_i^\alpha
\end{aligned}
\tag{21.11}
$$

from which it can be shown that the chemical potential of A is given by

$$
\mu_A = z W_{AB} y_B^\alpha y_B^\beta + z W_{CA} y_C^\alpha y_C^\beta + \tfrac{1}{2} \log_e (y_A^\alpha y_A^\beta)
\tag{21.12}
$$

with analogous relations for μ_B and μ_C.

Equation (21.11) is a free-energy functional with internal variables, the $y_i^{(j)}$, and has to be minimized subject to the mass balance constraints, for example,

$x_C = 0.5(y_C^\alpha + y_C^\beta)$, in order to obtain the equilibrium values of the sublattice mole fractions and the thermodynamic properties.

Similar equations can be used for the four-point cluster model discussed previously for binary alloys.

For both the two- and four-sublattice modeling equations, extra terms, involving ternary and/or quaternary parameters, could be added if required in the fitting to experimental results.

21.2 PHASE EQUILIBRIA

The most common type of binary phase diagram encountered in MS&E shows temperature as a function of composition at constant p, normally at 1 bar. With an additional component, one further variable must be fixed if a two-dimensional phase diagram is required for the ternary system. The phase diagrams are most frequently presented as isothermal sections at constant p (1 bar). Many such isothermal sections are required in order to gain a complete understanding of the temperature–composition relationships for a ternary system.

Equation (21.1) together with the use of either (21.7) (substitutional phases) or (21.12) (sublattice phases) may be solved to calculate the position of phase equilibrium between two preselected phases. We will illustrate the procedure by considering three examples.

Example 21.1 Miscibility Gap System
Assume as given the following RK parameters for the constituent binary systems:

$$L_0^{AB} = +20 \text{ kJ mol}^{-1}$$
$$L_1^{AB} = -5 \text{ kJ mol}^{-1}$$
$$L_0^{CA} = L_0^{BC} = +10 \text{ kJ mol}^{-1}$$

Calculate an isothermal section at 1000 K for the A–B–C system by using the RKM equation.

There is only a single phase to consider in this simple example. It is clear from Figure 21.3a that, at 1000 K, the A–B binary exhibits a miscibility gap while the other two binaries do not.

For a ternary alloy with an overall composition inside the miscibility gap, there are five unknowns, namely, $x_A^\alpha, x_A^{\alpha'}, x_B^\alpha, x_B^{\alpha'}$, and f, the fraction of the α-phase. The equilibrium values of the independent variables can be obtained from the solution of the following five equations:

$$\mu_A^\alpha(x_A^\alpha) = \mu_A^{\alpha'}(x_A^{\alpha'}) \tag{21.13}$$

$$\mu_B^\alpha(x_B^\alpha) = \mu_B^{\alpha'}(x_B^{\alpha'}) \tag{21.14}$$

$$\mu_C^\alpha(x_C^\alpha) = \mu_C^{\alpha'}(x_C^{\alpha'}) \tag{21.15}$$

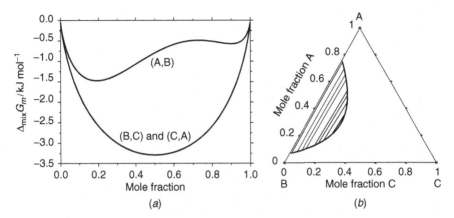

Figure 21.3 Binary $\Delta_{\text{mix}} G_m$ and calculated ternary isothermal section at 1000 K: (a) $\Delta_{\text{mix}} G_m$ for binaries; (b) isothermal section for ternary system.

$$x_A = f x_A^\alpha + (1 - f) x_A^{\alpha'} \tag{21.16}$$

$$x_B = f x_B^\alpha + (1 - f) x_B^{\alpha'} \tag{21.17}$$

where the chemical potentials are given by

$$\mu_i^\alpha = \mu_i^{\circ,\alpha} + RT \, \log_e x_i^\alpha + \mu_i^{\text{E},\alpha} \tag{21.18}$$

and $\mu_i^{\text{E},\alpha}$ is given by (21.7).

These equations may be solved at a fixed value of x_C and a fixed ratio of x_B/x_A, which guarantees a composition lying within the two-phase region. Using $x_C = 0.05$ and $x_B/x_A = 1.111$, the solution of these equations at 1000 K gives the solution for the independent variables:

$$x_A^\alpha = 0.09749 \qquad x_B^\alpha = 0.8412$$

$$x_A^{\alpha'} = 0.7119 \qquad x_B^{\alpha'} = 0.244 \tag{21.19}$$

$$f = 0.3449$$

The phase diagram shown in Figure 21.3b shows the complete isothermal section at 1000 K plotted on triangular coordinates (Gibbs triangle). An example of a real system which shows a phase diagram similar to this is the fcc (Au,Cu,Ni) phase in the Au–Cu–Ni system, with the miscibility gap extending from the binary Au–Ni.

Example 21.2 Liquid-Phase/Sublattice-Phase equilibrium

An isothermal section of a calculated phase diagram for the equilibrium between liquid and B₂ phases is shown in Figure 21.4. We wish to confirm this phase diagram by calculating the tie-line marked as l–b on the diagram.

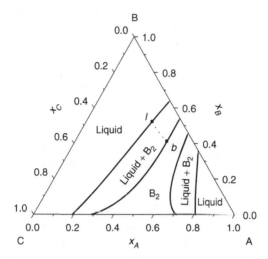

Figure 21.4 Isothermal phase diagram for system showing equilibrium between liquid phase and sublattice phase. One tie-line in the two-phase region is shown.

The isothermal section is for a temperature of 1050 K and the endpoints of the tie-line are given in Table 21.1.

The parameters used in the calculation of Figure 21.4, are as follows: The stabilities of the pure bcc components relative to the pure liquid components are given by

$$G(A, \text{bcc}) = G(B, \text{bcc}) = G(C, \text{bcc}) = -10 + 0.01T \text{ kJ mol}^{-1}$$

For the liquid solution phase, the binary RK parameters are

$$L_0^{AB} = -5 \text{ kJ mol}^{-1} \qquad L_0^{AC} = -4 \text{ kJ mol}^{-1} \qquad L_0^{BC} = 0$$

These may be used to calculate the ternary properties by using simplified forms of the RKM equation given in (21.7)—only the L_0 terms have to be considered.

For the two-sublattice B$_2$ (bcc) phase, the pair exchange energies are

$$W_{AB} = W_{BA} = -5 \text{ kJ mol}^{-1}$$

$$W_{AC} = W_{CA} = -4 \text{ kJ mol}^{-1}$$

$$W_{BC} = W_{CB} = +1.25 \text{ kJ mol}^{-1}$$

TABLE 21-1 Compositions of the Tie-Line 1-B in Figure 21.4

	x_A	x_B	x_C
Liquid	0.460173	0.403839	0.135988
B2	0.331947	0.533263	0.134789

These pair exchange parameters may be used with (21.12) and similar equations for μ_B and μ_C to calculate the chemical potentials in the B_2 phase relative to the pure bcc components.

We start by using, as the starting composition, the mean of the tie-line end-points. We will again use (21.13)–(21.17) but these must be supplemented since there are now seven independent variables, x_A^1, x_B^1, y_A', y_A'', y_B', y_B'', and f. The two extra constraints are

$$x_A(B2) = 0.5(y_A' + y_A'') \qquad (21.20)$$

$$x_B(B2) = 0.5(y_B' + y_B'') \qquad (21.21)$$

Solution of the seven equations yields the values for the seven independent variables. The calculated mole fractions are given in Table 21.1 and $f = 0.5$ since we took the average of the tie-line endpoints as a starting point in the calculation.

An example of a real system which shows features rather similar to the hypothetical one used in the calculation is the Au–Zn–Cd system. There are many other phases to be considered in this real system so that the simple features shown in our calculated isothermal section for the stable equilibria would be modified considerably due to the interference from the presence of other phases. Nevertheless, a metastable phase diagram which considered only the liquid and B_2 phases would be rather similar in the real and hypothetical system.

Example 21.3 A System with Binary Line Compound Phases

Figure 21.5 shows possible phase diagrams for a prototype ternary system where the binary Intermetallic Compounds (IMCs) exhibit a negligible range of homogeneity. This is a limiting case but, nevertheless, is found to occur quite frequently in real systems, particularly in nonmetallic systems. The lack of solution phases in this type of ternary makes phase diagram calculation rather straightforward; a computer solution is not required provided the thermodynamic properties of the IMCs are known.

We will illustrate a method for calculating a phase diagram which involves such line compounds. We assume that the Gibbs energies of all the phases are

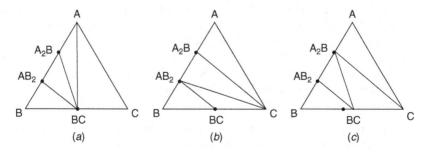

Figure 21.5 Possible isothermal sections for A–B–C system.

known with the following values of $\Delta_f G_m^\circ$ at 1000 K:

$$A_2B = -10.00 \text{ kJ mol}^{-1}$$
$$AB_2 = -9.67 \text{ kJ mol}^{-1}$$
$$BC = -9.00 \text{ kJ mol}^{-1}$$

The three possibilities for this phase diagram are shown in Figure 21.5. There it can be seen that the AB_2–BC join is common to all three possible phase diagrams. The other two joins are seen to vary between where both involve BC (Fig. 21.5a), both involve C (Fig. 21.5b), and one involves BC and one C (Fig.21.5c). We may determine which is the correct phase diagram by considering the reaction equations

$$3BC + A_2B = C + 2AB_2 \tag{21.22}$$
$$3C + 2A = C + A_2B \tag{21.23}$$

The values for $\Delta_r G^\circ$ for these two reactions are

$$2(-9.67) - [3(-9) + (-10)] = +27.66$$
$$(-10) - [(-9)] = -1$$

from which we conclude that Figure 21.5c represents the stable phase diagram in this case.

EXERCISES

21.1 Derive (21.12).

21.2 Confirm the results given in (21.19).

21.3 Confirm the results given in Table 21.1.

21.4 Derive the equation for the Kohler interpolation for the case where all three binaries are strictly regular. (*Hint:* The ratio of two components is the same in the ternary and binary alloys.) Compare the estimated values for G^E in ternary solutions of composition ($x_A = 0.333$, $x_B = 0.333$) and ($x_A = 0.8$, $x_B = 0.1$) when the RKM and Kohler methods are used for strictly regular solutions. Which extrapolation method do you think is the better?

21.5 (a) There is one intermediate phase in the In–P system, InP, two IMCs in the In–Co system, $CoIn_3$ and $CoIn_2$, and three IMCs in the Co–P system, PCo_2, PCo, and P_3Co.

The enthalpies of formation of these IMCs at 298.15 K are InP (-18.85); $CoIn_3$ (-0.975), $CoIn_2$ (-2.17); PCo_2 (-21.77), PCo (-39.55), and P_3Co (-18.1), all expressed in term of kilojoules per mole of atoms.

On the basis of these data, calculate the composition isotherm for the In–Co–P system at 298.15 K. What assumption do you have to make in order to calculate this diagram?

(b) Suppose that you need to make an electrical contact to an InP alloy. Would you recommend the use of Co at room temperature? What about at 373 K? Why?

21.6 Ni and Si form three IMCs, Ni_3Si, Ni_5Si_2, and Ni_2Si. Si and C form only one, SiC. Calculate the ternary phase diagram of Ni–Si–C at 600°C.

The standard Gibbs energies of formation of these phases are Ni_2Si (-142.7), Ni_5Si_2 (-315.0), Ni_3Si (-160.0), and SiC (-66.9), all expressed in term of kilojoules per mole of formula.

21.7 For a ternary A–B–C system, there exist three binary intermetallic phases, AB, BC, and AC.

The standard Gibbs energies of formation of these three compounds at 1000 K are $A_{0.5}B_{0.5}$ (-10), $A_{0.5}C_{0.5}$ (-20), and $B_{0.5}C_{0.5}$ (-5), all expressed in terms of kilojoules per mole of atoms.

(a) Calculate the 1000-K isotherm assuming there are mutual solubilities in pure A, B, and C and the three binary compounds are negligible.

(b) Calculate and plot an isothermal field–density phase diagram using μ_C/RT as a function of $x_B/(x_B + x_A)$ at 1000 K. Assume that there are no mutual solubilities between any of the phases. Draw another (schematic) diagram where finite solubilities between these phases is assumed.

22 Generalized Phase Diagrams for Ternary Systems

In discussing unary and binary systems, we have already met some different ways of graphically presenting the results from equilibrium calculations in the form of two-dimensional phase diagrams.

(A) *Unary* For unary systems, phase diagrams can be plotted in two dimensions without any constraints. This can be understood from the Gibbs–Duhem equation for a unary system:

$$S\, dT - V\, dp + n\, d\mu = 0$$

or, in molar form,

$$S_m\, dT - V_m\, dp + d\mu = 0 \tag{22.1}$$

from which it can be seen that, for given values of any two of the field variables, the third is automatically adjusted.

This leads to the following classification for unary, two-dimensional phase diagrams:

- (i) field–field phase diagrams, for example, p versus T
- (ii) field–density phase diagrams, for example, p versus S_m but not involving conjugate variables such as T versus S_m
- (iii) density–density phase diagrams, for example, S_m versus V_m

(B) *Binary* For binary systems, it is necessary to fix one independent variable in order to be able to plot a two-dimensional phase diagram. Most binary phase diagrams used in MS&E refer to a constant total pressure of 1 bar and, under these constant-pressure conditions, the Gibbs–Duhem equation is

$$S\, dT + n_A\, d\mu_A + n_B\, d\mu_B = 0$$

or, in molar form,

$$S_m\, dT + (1 - x_B)\, d\mu_A + x_B\, d\mu_B = 0 \tag{22.2}$$

Materials Thermodynamics. By Y. Austin Chang and W. Alan Oates
Copyright 2010 John Wiley & Sons, Inc.

from which it can be seen that, if two field variables are fixed, the third is automatically adjusted. As a result, constant-total-pressure two-dimensional phase diagrams can be classified in the same way as for unary systems:

  (i) field–field phase diagrams, for example, T versus μ_B

  (ii) field–density phase diagrams, for example, T versus x_B and μ_B versus S_m but not involving conjugate variables such as T versus S_m

  (iii) density–density phase diagrams, for example, S_m versus x_B

(C) *Ternary* The situation for ternary systems may be understood by again considering the constant-total-pressure Gibbs–Duhem equation:

$$S\,dT + n_A\,d\mu_A + n_B\,d\mu_B + n_C\,d\mu_C = 0$$

This equation can be manipulated in different ways to convert the extensive quantity to a density:

$$\frac{S}{n_C}\,dT + \frac{n_A}{n_C}\,d\mu_A + \frac{n_B}{n_C}\,d\mu_B + d\mu_C = 0$$

$$\frac{S}{n_A + n_B}\,dT + \frac{n_A}{n_A + n_B}\,d\mu_A + \frac{n_B}{n_A + n_B}\,d\mu_B + \frac{n_C}{n_A + n_B}\,d\mu_C = 0$$

This last equation can be written more compactly as

$$S_n\,dT + (1 - \xi)\,d\mu_A + \xi\,d\mu_B + r\,d\mu_C = 0 \tag{22.3}$$

where

$$S_n = \frac{S}{n_A + n_B} \qquad \xi = \frac{n_B}{n_A + n_B} \qquad r = \frac{n_C}{n_A + n_B}$$

The use of ξ and r takes advantage of the fact that there are only two independent composition variables ($x_A + x_B + x_C = 1$).

It is clear that, for ternaries, it is now necessary to fix two properties before a two-dimensional phase diagram can be drawn. Constant total pressure is one obvious choice with a constant field or density being the other obvious choice. However, another possibility exists and we first discuss this before returning to a classification of the different possible phase diagrams for ternary systems.

Special Constraints in Ternary Systems. A situation with no counterpart in binary alloys may occur when one of the three alloying elements diffuses much more rapidly than the other two, for example, when one of the elements dissolves interstitially (e.g., H, C, N). The best known example of this, which we will consider, occurs during the transformation of austenite (γ) to ferrite (α) in Fe–C–X alloys, where X is a substitutional alloying element. The interface will be mobile during an actual transformation but, since we are concerned only with thermodynamics, we will consider the situation where the interface is immobile but where the Fe/X ratio remains constant.

In the case of complete equilibrium (all species are considered mobile), we can use $\mu_i^\alpha = \mu_i^\gamma$ for all three elements as the condition of equilibrium. At constant total p, a single phase in the ternary alloy has three degrees of freedom and it is therefore necessary to fix another variable if we wish to plot a two-dimensional phase diagram. The most obvious one to fix is T. Figure 22.1 shows a schematic complete equilibrium field–density phase diagram at high Fe concentrations (the dashed curves). If we start with the α-phase and increase the carbon potential, the α will first transform into the γ-phase of a different Fe/X ratio. Eventually, at the completion of the transformation, the γ-phase will have the same Fe/X ratio as the original.

The constrained equilibrium situation is quite different. First, because $(\text{Fe/X})^\alpha = (\text{Fe/X})^\gamma$ at all times, the system acts like a pseudobinary and, second, we can no longer use $\mu_i^\alpha = \mu_i^\gamma$ for all three components but only for C, $\mu_C^\alpha = \mu_C^\gamma$. We may obtain another constraining condition for the chemical potentials via the use of a Lagrangian multiplier, λ, for the mass balance constraint $N_{Fe}^\alpha N_X^\gamma - N_{Fe}^\gamma N_X^\alpha = 0$. The Lagrangian for the α/γ is given by

$$\mathcal{L} = G^\alpha + G^\gamma + \lambda(N_{Fe}^\alpha N_X^\gamma - N_{Fe}^\gamma N_X^\alpha)$$

and its minimization gives

$$\frac{\partial \mathcal{L}}{\partial N_{Fe}^\alpha} = \mu_{Fe}^\alpha + N_X^\gamma \lambda = 0$$

$$\frac{\partial \mathcal{L}}{\partial N_{Fe}^\alpha} = \mu_X^\alpha - N_{Fe}^\gamma \lambda = 0$$

$$\frac{\partial \mathcal{L}}{\partial N_{Fe}^\alpha} = \mu_{Fe}^\gamma - N_X^\alpha \lambda = 0$$

$$\frac{\partial \mathcal{L}}{\partial N_{Fe}^\alpha} = \mu_X^\gamma + N_{Fe}^\alpha \lambda = 0$$

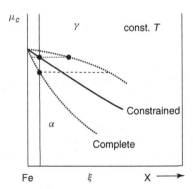

Figure 22.1 Schematic μ_C–ξ isotherm for Fe–C–X system. Both complete (dashed curves) and constrained (solid curve) equilibrium phase diagrams are shown.

The λ may be eliminated from these equations to give the equilibrium relation between the chemical potentials for this constrained equilibrium:

$$(1 - \xi)\mu_{Fe}^{\alpha} + \xi\mu_X^{\alpha} = (1 - \xi)\mu_{Fe}^{\gamma} + \xi\mu_X^{\gamma} \tag{22.4}$$

This last equation can be written as

$$G_n^{\alpha} = G_n^{\beta} \tag{22.5}$$

where the subscript refers to the unit amount of the immobile (Fe and X) components.

Another, rather similar, constraint may also occur in a ternary system. This can be seen by referring to Figure 22.2, which shows an isothermal section at 1600 K for the FeO–Fe$_2$O$_3$–SiO$_2$ system. It can be seen that the liquid phase can be saturated with either metallic or oxide phases. The metallic phase is not pure Fe but is an Fe–Si alloy. Following a similar argument to that given above for the constrained equilibrium in the Fe–X–C system, we can see that the equilibrium condition along this metal-saturated boundary, with pure Fe and pure Si as reference states, is

$$(1 - \xi)\mu_{Fe} + \xi\mu_{Si} = 0 \tag{22.6}$$

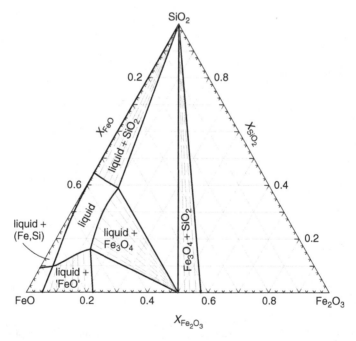

Figure 22.2   Isothermal section for FeO–Fe$_2$O$_3$–SiO$_2$ system at 1600 K.

We can now return to the classification of two-dimensional, constant-total-pressure phase diagrams for ternary systems:

(a) Field–field phase diagrams
    (i) at a constant field, for example, μ_A versus μ_B at constant T
    (ii) at a constant density, for example, T versus μ_C at constant ξ
    (iii) under a special constraint

(b) Field–density phase diagrams
    (i) at a constant field, for example, T versus ξ at constant μ_C
    (ii) at a constant density, for example, T versus r at constant ξ
    (iii) under a special constraint

(c) Density–density phase diagrams
    (i) at a constant field, for example, r versus ξ at constant T
    (ii) at a constant density, for example, V_m versus r at constant ξ
    (iii) under special constraint

Attention to the topological similarity of the three types of phase diagrams in one-, two-, and three-component systems should be emphasized. This is apparent in Figure 22.3, where some field–field phase diagrams are shown. All three figures show triple points with three lines emanating from the triple point. These lines can have a maximum (congruent point) or a minimum or may terminate at a (critical) point.

The topological similarity between unary, binary, and ternary systems is also found for field–density and density–density phase diagrams when the appropriate number of variables are held constant.

Some examples of these three types of two–dimensional phase diagrams for ternary systems are discussed in more detail below.

Example 22.1 μ_A **versus** μ_B **at constant** T
Figure 22.4a shows a plot of μ_{O_2} versus μ_{S_2} for the Fe–S–O system at 1000 K. As in any field–field phase diagram, this diagram is seen to be divided into

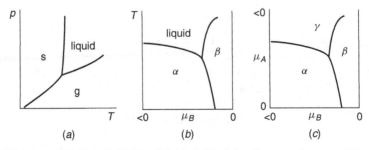

Figure 22.3 Topological similarity of field–field phase diagrams for unary, binary, and ternary systems: (a) unary system; (b) binary system, at constant p; (c) ternary system, at constant p, T.

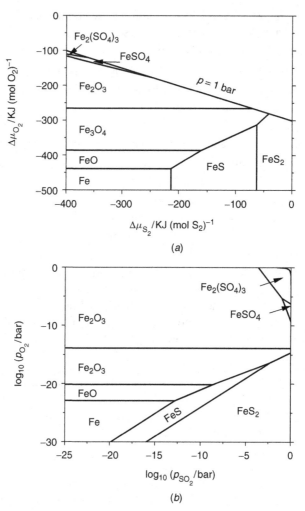

Figure 22.4 Isothermal sections at 1000 K for Fe–S–O system: (a) μ_{O_2} versus μ_{S_2}; (b) $\log_{10} p_{O_2}$ versus $\log_{10} p_{SO_2}$.

the various *domains of stability* for the various condensed phases and hence the name often given to this type of phase diagram is predominance area diagram.

In the diagram shown in Figure 22.4a the horizontal lines refer to simple oxidation reactions, for example,

$$2Fe(s) + O_2(g) = 2FeO(s)$$

while the vertical line refers to a simple sulfidation reactions, for example,

$$2Fe(s) + S_2(g) = 2FeS(s)$$

The sloping lines refer to reactions which involve both O_2 and S_2, for example,

$$FeO(s) + S_2(g) = FeS(s) + O_2(g)$$

The diagram axes selected for Figure 22.4a are not necessarily the most convenient ones to use in a practical application. It is straightforward to change to a more suitable pair of variables. These will be related to the two used in Figure 22.4a. Consider, for example, the chemical reaction equations

$$2SO_2 + O_2 = 2SO_3 \tag{22.7}$$

$$4SO_3 + S_2 = 6SO_2 \tag{22.8}$$

$$S_2 + 2O_2 = 2SO_2 \tag{22.9}$$

It can seen that the sulfur potential can be specified through $(p_{SO_2})^6/(p_{SO_3})^4$ or through $(p_{SO_2})/p_{O_2})^2$ and the oxygen potential through $(p_{SO_3}/p_{SO_2})^2$. If we were to select $\log_{10} p_{SO_2}$ as one axis, then we have several possible choices for the other axis (e.g., $\log_{10} p_{O_2}$). A phase diagram equivalent to Figure 22.4a is shown in Figure 22.4b.

Example 22.2 $T-\mu_i$ under Special Constraint
In the $FeO-Fe_2O_3-SiO_2$ system, an isothermal section for which was shown in Figure 22.2, oxygen potentials will vary as a function of composition across the liquid phase region. If we wish to introduce temperature as a variable, then it will be necessary to fix a different variable if a two-dimensional phase diagram is still required. One possibility is to fix the oxygen potential or partial pressure and show the phase diagram, for example, at 1 bar of air. Another is to show the temperature variation of the oxygen potentials along the alloy-saturated boundary which, thermodynamically, corresponds with the chemical potential condition given previously in (22.6). Under this constraint, the $T-\mu_{O_2}$ phase diagram shown in Figure 22.5 is obtained.

A brief explanation of this phase diagram is warranted. The curve shown at the highest oxygen potentials in Figure 22.5 corresponds with the Fe/FeO equilibrium. At oxygen potentials higher than those given by this curve, we will no longer have metal saturation with the higher oxides of Fe becoming stable. The curve shown at the lowest oxygen potentials in the diagram corresponds with the Si/SiO_2 equilibrium. At oxygen potentials lower than those given by this curve, only Si can exist. In the region between the upper and lower bounds set by the Fe/FeO and Si/SiO_2 curves metal saturation must correspond with alloys of intermediate compositions: Si-rich alloys at low oxygen potentials and Fe-rich alloys as the Fe/FeO curve is approached. In other words, the domains of stability for the different phases in the intermediate region must resemble those found in a $T-\mu_{Si}$ phase diagram for the Fe−Si system. These domains of stability for the different phases are indicated on the areas in Figure 22.5. It is also possible to insert lines for different compositions of both the metallic and oxide phases on

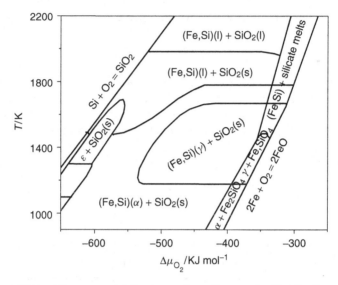

Figure 22.5 Alloy-saturated $T - \Delta\mu_{O_2}$ phase diagram for Fe–Si–O system.

this phase diagram. In the liquid-phase region for the liquid alloys, lines for the concentration of O as well as those for Si could also be inserted. Such lines are useful for indicating the deoxidation behavior of molten steel by Si.

Example 22.3 Field–Density at a Constant Field

A 773-K isotherm for the Fe–Cr–O ternary system is shown in Figure 22.6a. Several line compounds are present in this system, that is, Fe_3O_4 and Fe_2O_3 in the binary Fe–O and Cr_2O_3 in the binary Cr–O. Solution phases of (Fe,Cr) are present in the binary Fe–Cr system and in $(Fe,Cr)_2O_3$ in the pseudobinary Fe_2O_3–Cr_2O_3 system. A ternary spinel phase $(Fe,Cr)_3O_4$ is found along the 57.1 at % O join, but this phase is stable only in the ternary composition region, not in the binary Fe–O and Cr–O systems.

The type of phase diagram presentation given in Figure 22.6a can be supplemented by a field–density isotherm like that presented in Figure 22.6b.

The relative utility of these different forms of presentation will vary: For those interested in open systems, Figure 22.6b is the more useful method of presentation, while the reverse is true for those interested in closed systems.

Example 22.4 Three-Dimensional Phase Diagrams

Three-dimensional phase diagrams overcome the constraints imposed by plotting two-dimensional diagrams. A $\mu_A - \mu_B - \mu_C$ isotherm for the Fe–S–O system is shown in Figure 22.7. While such phase diagrams have a role, it is apparent from this figure that it is difficult to obtain quantitative information about equilibria occurring in the center of the figure.

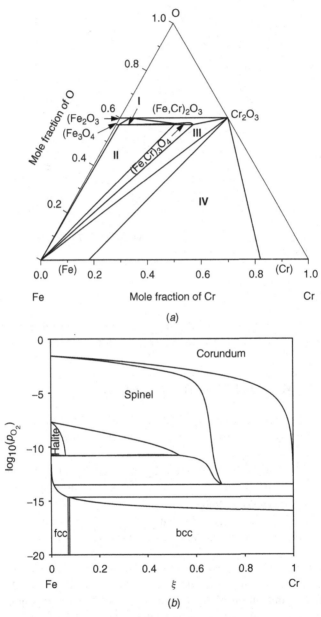

Figure 22.6 Phase diagrams for Fe–Cr–O system: (*a*) density–density isothermal section at 773 K; (*b*) field–density isothermal section at 1573 K.

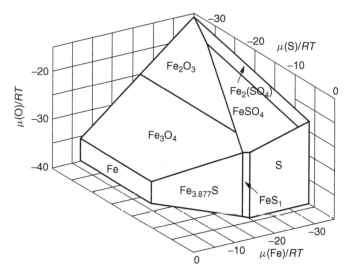

Figure 22.7 Isothermal section of Fe–S–O system at 1973 K. *Source:* H. Yokokawa, *J. Phase Equilib.*, **20**, pp. 258–287, 1999.

22.1 SYSTEM DEFINITION

Here we will further develop the discussion begun in Chapter 20 concerning the importance of using a good system definition before performing thermodynamic calculations.

Example 22.5 Melting of Ti in Al_2O_3 Crucibles

(i) Thermodynamicist A simply used the data presented on an Ellingham diagram. The system definition is poor and, as a result, an incorrect answer was obtained.

(ii) Thermodynamicist B defined the system better and obtained the correct answer by appreciating that a low O concentration in Ti(l) requires an extremely low oxygen potential.

(iii) Thermodynamicist C realized that it is necessary to consider all possible phases occurring in the ternary Ti–Al–O system.

If the system is assumed to be completely closed at 2200 K (no loss of Ti, Al, or O from the system), then any contact between Ti and Al_2O_3 must result in some final composition along the heavy dashed curve in Figure 22.8. The heavy solid curve in that figure represents a crude guess of the location of the liquid compositions at this temperature. It can be seen that the intersection of this phase boundary curve with the dashed mass balance constraint curve will result in a liquid phase which is high in both O and Al. The oxygen potential of this $(Ti,Al,O)(l)–Al_2O_3$ system will be considerably higher than that required for

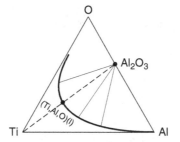

Figure 22.8 Isothermal section for Ti–Al–O system at 2200 K.

the preparation of O–free Ti (a guess of around -600 instead of approximately -1100 kJ $(\text{mol } O_2)^{-1}$)

Thermodynamicist C is also aware that this conclusion may be faulty because of the closed-system assumption. He or she is aware of the possibility of volatile species being present and carries out a back-of-the-envelope calculation for the Al–O system by constructing an Ellingham diagram as shown in Figure 22.9. It can be seen that, at 2200 K and an oxygen potential of around -600 kJ (mol $O_2)^{-1}$, the partial pressure of $Al_2O(g)$ could be significant so that volatile species should be considered in a full thermodynamic analysis.

These considerations by thermodynamicist C emphasize not only the importance of having a good system definition but also the difficulties associated with trying to arrive at this aim. In high-temperature systems like the one discussed,

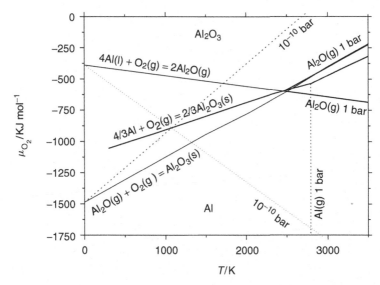

Figure 22.9 Ellingham diagram which takes into account some vapor species for Al–O system.

the calculation may be very difficult qualitatively and often impossible quantitatively because of the lack of the necessary thermochemical data for all the species in the defined system.

EXERCISES

22.1 (a) Calculate the 1300-K isotherm for the section Cu–Co–CoO–CuO. Ignore any equilibria on the oxygen-rich side of the CuO–CoO section and assume that there is negligible solubility between the following phases: Cu and Co, Co and CoO, Cu and Cu_2O, Cu_2O and CuO, CuO and CoO.

  (b) Calculate the values of log p_{O_2} at the phase equilibria.

22.2 For the unary, binary, and ternary field–field phase diagrams given in Figure 22.3, sketch the corresponding field–density and density–density phase diagrams.

APPENDIX A
Some Linearized Standard Gibbs Energies of Formation

Substance	Temperature Range/K	$\Delta_f\ G_m^\circ/\text{J mol}^{-1}$
$CO_2(g)$	298.15–2500	$-394{,}886-0.77T$
$CO(g)$	298.15–2500	$-113{,}262-86.37T$
$CH_4(g)$	298.15–2500	$-86{,}846+107.77T$
$H_2O(g)$	298.15–1500	$-246{,}490+54.8T$
$SO_2(g)$	298.15–2500	$362{,}420+72.4T$
$SO_3(g)$	298.15–2500	$-456{,}980+161.8T$
$SiO_2(\text{quartz})$	298.15–1687	$-906.71+0.176T$
$SiO_2(\text{quartz})$	1687–1983	$-947.29+0.199T$
$Fe_{0.95}O(s)$	298.15–1184	$-263{,}376+64.9T$
$Fe_{0.95}O(s)$	1184–1600	$-263{,}332+64.8T$
$Cu_2O(s)$	298.15–1357	$-169{,}337+73.4T$
$CuO(s)$	298.15–1357	$-152{,}728+86.0T$
$CoO(s)$	298.15–2000	$-234{,}689.5+71.1T$
$Al_2O_3(s)$	298–933.47	$-167{,}5430+313T$
$Al_2O_3(s)$	933.47–2323	$-1{,}682{,}880+324T$

Materials Thermodynamics. By Y. Austin Chang and W. Alan Oates
Copyright 2010 John Wiley & Sons, Inc.

APPENDIX B
Some Useful Calculus

RELATIONS BETWEEN PARTIAL DERIVATIVES

For simple unary systems (those which undergo only $p-V$ work), the principal thermodynamic functions are p, V, T, U, S, A, H, and G. More than 300 partial derivatives can be formed from these 8 variables and more than 5×10^8 formulas can be formed from the derivatives.

Some rationalization of the most important formulas can be obtained by expressing all other derivatives in terms of the standard observable quantities, $(\partial V / \partial T)_p$, $(\partial V / \partial p)_T$, and $(\partial H / \partial T)_p$.

The following four methods can be used for this purpose:

1. The standard relation between a total differential and the partial derivatives for a function $z = z(x, y)$ is

$$dz = \left(\frac{\partial z}{\partial x}\right)_y dx + \left(\frac{\partial z}{\partial y}\right)_x dy \tag{B-1}$$

This relation allows us to use a Gibbs equation like $dG = V\,dp - S\,dT$ to obtain the relations

$$\left(\frac{\partial G}{\partial p}\right)_T = V \tag{B-2}$$

$$\left(\frac{\partial G}{\partial T}\right)_p = -S \tag{B-3}$$

2. The relation between ordinary derivatives,

$$\frac{dy}{dx} = \frac{dy/du}{dx/du} \tag{B-4}$$

Materials Thermodynamics. By Y. Austin Chang and W. Alan Oates
Copyright 2010 John Wiley & Sons, Inc.

also applies to partial derivatives, for example,

$$\left(\frac{\partial y}{\partial x}\right)_z = \frac{(\partial y/\partial u)_z}{(\partial x/\partial u)_z} \qquad \text{(B-5)}$$

We can use this equation to see, for example, that

$$\left(\frac{\partial A}{\partial p}\right)_T = \frac{(\partial A/\partial V)_T}{(\partial p/\partial V)_T} \qquad \text{(B-6)}$$

When this is used in conjunction with a Gibbs equation, for example, $dA = -p\,dV - S\,dT$, from which

$$\left(\frac{\partial A}{\partial V}\right)_T = -p \qquad \text{(B-7)}$$

then we obtain

$$\left(\frac{\partial A}{\partial p}\right)_T = -p\left(\frac{\partial V}{\partial p}\right)_T \qquad \text{(B-8)}$$

3. Equation (B.1) can also be used to obtain another relation which is often used for changing variables. If z is maintained constant in (B.1), then

$$0 = \left(\frac{\partial z}{\partial x}\right)_y + \left(\frac{\partial z}{\partial y}\right)_x \left(\frac{\partial y}{\partial x}\right)_z \qquad \text{(B-9)}$$

or

$$\left(\frac{\partial z}{\partial y}\right)_x \left(\frac{\partial y}{\partial x}\right)_z \left(\frac{\partial x}{\partial z}\right)_y = -1 \qquad \text{(B-10)}$$

As a thermodynamic example, we can use this equation to see that

$$\left(\frac{\partial p}{\partial T}\right)_V \left(\frac{\partial T}{\partial V}\right)_p \left(\frac{\partial V}{\partial p}\right)_T = -1 \qquad \text{(B-11)}$$

4. There is an important theorem for reciprocal relations between the second partial derivatives. Using (B.1), this theorem is given by

$$\left(\frac{\partial}{\partial y}\left(\frac{\partial z}{\partial x}\right)_y\right)_x = \left(\frac{\partial}{\partial x}\left(\frac{\partial z}{\partial y}\right)_x\right)_y \qquad \text{(B-12)}$$

The so-called Maxwell relations are obtained by applying this theorem to the Gibbs equations. For example, for the Gibbs equation $dG = V\,dp - S\,dT$, its application leads to

$$\left(\frac{\partial V}{\partial T}\right)_p = -\left(\frac{\partial S}{\partial p}\right)_T \qquad \text{(B-13)}$$

Example B.1 As an illustration of the use of these four methods, we will obtain an expression for $(\partial G/\partial T)_V$ in terms of the standard measurable partial derivatives.

From $dG = V\,dp - S\,dT$, we see that

$$\left(\frac{\partial G}{\partial T}\right)_p = -S \qquad \left(\frac{\partial G}{\partial p}\right)_T = V \tag{B-14}$$

and then, using (B.1) for $G = G(p, T)$,

$$dG = \left(\frac{\partial G}{\partial T}\right)_p dT + \left(\frac{\partial G}{\partial p}\right)_T dp \tag{B-15}$$

we can now obtain the desired partial derivative:

$$\left(\frac{\partial G}{\partial T}\right)_V = \left(\frac{\partial G}{\partial T}\right)_p \left(\frac{\partial T}{\partial T}\right)_V + \left(\frac{\partial G}{\partial p}\right)_T \left(\frac{\partial p}{\partial T}\right)_V \tag{B-16}$$

$$= -S + V\left(\frac{\partial p}{\partial T}\right)_V \tag{B-17}$$

$$= -S - V\frac{(\partial V/\partial T)_p}{(\partial V/\partial p)_T} \tag{B-18}$$

EULER'S THEOREM FOR HOMOGENEOUS FUNCTIONS

If we can write

$$f(\alpha x_1, \ldots, \alpha x_N) = \alpha^n f(x_1, \ldots, x_N) \tag{B-19}$$

then $f(x_1, \ldots, x_N)$ is said to be a homogeneous function of degree n.

As an example, consider the function

$$f(x, y) = \frac{x}{x^2 + y^2} \tag{B-20}$$

Then

$$f(\alpha x, \alpha y) = \frac{\alpha x}{\alpha^2 x^2 + \alpha^2 y^2} = \alpha^{-1} f(x, y) \tag{B-21}$$

so that this function is homogeneous of degree -1.

In thermodynamics we deal with homogeneous functions of degree 1. For these functions, Euler's theorem states that

$$\sum_i^N x_i \frac{\partial f}{\partial x_i} = f(x_1, \ldots, x_N) \tag{B-22}$$

Its proof is given in standard texts on advanced calculus. We will illustrate the application of the theorem to some thermodynamic situations.

(i) For volume, $V(p, T, n_1, \ldots, n_N)$, only the amounts enter into the theorem:

$$V(p, T, \alpha n_a, \ldots, \alpha n_N) = \alpha V(p, T, n_1, \ldots, n_N) \qquad \text{(B-23)}$$

and so

$$V(p, T, n_1, \ldots, n_N) = \sum_i^N n_i \left(\frac{\partial V}{\partial n_i}\right)_{p, T\{n_j\}} \qquad \text{(B-24)}$$

$$= \sum_i^N n_i V_i \qquad \text{(B-25)}$$

(ii) The Helmholtz energy for a unary system is homogeneous of degree 1 in the volume and the amount of substance. Application of Euler's theorem in this case gives

$$A(T, V, N) = V \left(\frac{\partial A}{\partial V}\right)_{T,N} + N \left(\frac{\partial A}{\partial N}\right)_{T,V} \qquad \text{(B-26)}$$

$$= -pV + \mu N \qquad \text{(B-27)}$$

(iii) The Gibbs energy for a multicomponent system is homogeneous of degree 1 in the amounts of substances. Application of Euler's theorem gives

$$G(T, p, n_1, \ldots, n_N) = \sum_i^N n_i \left(\frac{\partial G}{\partial n_i}\right)_{T,p} \qquad \text{(B-28)}$$

$$= \sum_i^N n_i \mu_i \qquad \text{(B-29)}$$

STATIONARY VALUES UNDER CONSTRAINTS

A situation which is often encountered in thermodynamics is the need to minimize a function subject to certain other constraints which arise from the fact that some or all of the other variables are not truly independent. We can illustrate this by considering a simple example.

Example B.2 Find the minimum of

$$z = (x - 2)^2 + (y - 3)^2$$

subject to the constraint

$$4x + y = 3$$

The unconstrained minimum of the function, when x and y are independent, can be found from $dz = 0$:

$$dz = \left(\frac{\partial z}{\partial x}\right)_y dx + \left(\frac{\partial z}{\partial y}\right)_x dy \qquad \text{(B-30)}$$

$$0 = \left(\frac{\partial z}{\partial x}\right)_y \qquad \text{(B-31)}$$

$$0 = \left(\frac{\partial z}{\partial y}\right)_x \qquad \text{(B-32)}$$

In our example,

$$\text{Min } z = (x - 2)^2 + (y - 3)^2 \qquad \text{(B-33)}$$

$$0 = \left(\frac{\partial z}{\partial y}\right)_x = 2x - 4 \qquad \text{(B-34)}$$

$$0 = \left(\frac{\partial z}{\partial x}\right)_y = 2y - 6 \qquad \text{(B-35)}$$

that is, the minimum for this function, which is shown in Figure B.1a, occurs at $x = 2, y = 3$.

The effect of the constraint is to shift the attainable minimum in the function, as can be seen from the intersection of the function and the constraining relation in Figure B.1b.

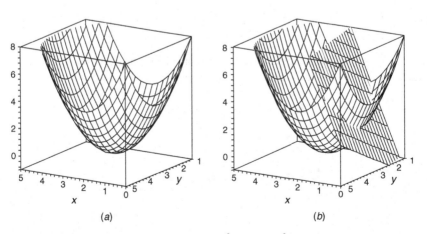

Figure B.1 Minimum in function $z = (x - 2)^2 + (y - 3)^2$: ($a$) no constraints; ($b$) under constraint $4x + y = 3$.

It is possible to solve this constrained minimization problem by incorporating the constraint equation into the function equation (B.33). If we substitute for y, we obtain

$$z = (x - 2)^2 + (-4x)^2 \qquad \text{(B-36)}$$

and we may now obtain the stationary point from $(dz/dx) = 0$:

$$\frac{dz}{dx} = 34x - 4 = 0 \qquad \text{(B-37)}$$

$$x = \tfrac{4}{34} = 0.118 \qquad \text{(B-38)}$$

The value of $y = 2.529$ at the stationary point can be obtained by substituting for x in the constraint equation.

An alternative method for solving this problem is to use *Lagrange's method of undetermined multipliers*. This method, which is superior to the substitution method when there are several constraints or when some constraints are nonlinear, is the one which has been used throughout this text.

The total differentials for both the original function $z = z(x, y)$ and the constraining function, which we can write as $g(x, y) = 0$, are zero at the constrained minimum, that is, $dz = 0$ and $dg = 0$. But neither $(\partial z/\partial x)_y$ nor $(\partial z/\partial y)_x$ are zero at this minimum because we now have only one independent variable. But if we define a new Lagrangian function $\mathcal{L}$ as

$$\mathcal{L} = z + \lambda g \qquad \text{(B-39)}$$

then

$$d\mathcal{L} = dz + \lambda\, dg \qquad \text{(B-40)}$$

$$= \left(\frac{\partial z}{\partial x} + \lambda \frac{\partial g}{\partial x} \right) + \left(\frac{\partial z}{\partial y} + \lambda \frac{\partial g}{\partial y} \right) \qquad \text{(B-41)}$$

The unconstrained Lagrangian is now zero at the stationary point so that the two bracketed terms are individually also zero. Applying this to our example,

$$\mathcal{L} = (x - 2)^2 + (y - 3)^2 + \lambda(4x + y - 3) \qquad \text{(B-42)}$$

$$\frac{\partial \mathcal{L}}{\partial x} = 2x - 4 + 4\lambda = 0 \qquad \text{(B-43)}$$

$$\frac{\partial \mathcal{L}}{\partial y} = 2y - 6 + \lambda = 0 \qquad \text{(B-44)}$$

Elimination of the λ gives

$$x - 4y + 10 = 0 \tag{B-45}$$

which can be combined with the constraining equation $4x + y - 3 = 0$ to give the same values for x and y at the constrained minimum as obtained previously by the substitution method.

Index

Materials Thermodynamics. By Y. Austin Chang and W. Alan Oates
Copyright 2010 John Wiley & Sons, Inc.

Printed in the United States
By Bookmasters